S0-APQ-346

THE WORLD OF MACHINES

Curriculum Consultants

Dr. Arnold L. Willems
Associate Professor of Curriculum and Instruction
The University of Wyoming

Dr. Gerald W. Thompson
Associate Professor
Social Studies Education
Old Dominion University

Dr. Dale Rice
Associate Professor
Department of Elementary and Early Childhood Education
University of South Alabama

Dr. Fred Finley
Assistant Professor of Science Education
University of Wisconsin

Subject Area Consultants

Astronomy
Robert Burnham
Associate Editor
Astronomy Magazine and *Odyssey* Magazine

Geology
Dr. Norman P. Lasca
Professor of Geology
University of Wisconsin — Milwaukee

Oceanography
William MacLeish
Editor
Oceanus Magazine

Paleontology
Linda West
Dinosaur National Monument
Jensen, Utah

Physiology
Kirk Hogan, M.D.
Madison, Wisconsin

Sociology/Anthropology
Dr. Arnold Willems
Associate Professor of Curriculum and Instruction
College of Education
University of Wyoming

Technology
Dr. Robert T. Balmer
Professor of Mechanical Engineering
University of Wisconsin — Milwaukee

Transportation
James A. Knowles
Division of Transportation
Smithsonian Institution

Irving Birnbaum
Air and Space Museum
Smithsonian Institution

Donald Berkebile
Division of Transportation
Smithsonian Institution

Zoology
Dr. Carroll R. Norden
Professor of Zoology
University of Wisconsin —
Milwaukee

Managing editor
Patricia Daniels

Editors
Herta Breiter
Darlene Shinozaki Kuhnke
Patricia Laughlin
Norman Mysliwiec

Designers
Faulkner/Marks
Jane Palecek

Artists
Jim Bamber
Brian Cody
Dick Eastland
Philip Emms
Dan Escott
Elizabeth Graham-Yool
Colin Hawkins
Eric Jewell
Ben Manchipp
Stephanie Manchipp
Barry Salter
Michael Welply

First published by Macmillan Publishers Limited, 1979

Library of Congress Number: 80-22980
1 2 3 4 5 6 7 8 9 84 83 82 81
Printed and bound in the United States of America.

Library of Congress Cataloging in Publication Data
Main entry under title:

Let's discover the world of machines.

(Let's discover ;)
Bibliography: p. 69
Includes index.

SUMMARY: A reference book dealing with simple machines, machines in the home, transportation and power machines, and large machinery found on the farm, in construction, and in factories.
1. Machinery — Juvenile literature.
[1. Machinery] I. Title: World of machines.
II. Series.
AG6.L43 [TJ147] 031s [621.8]
ISBN 0-8172-1756-8 80-22980

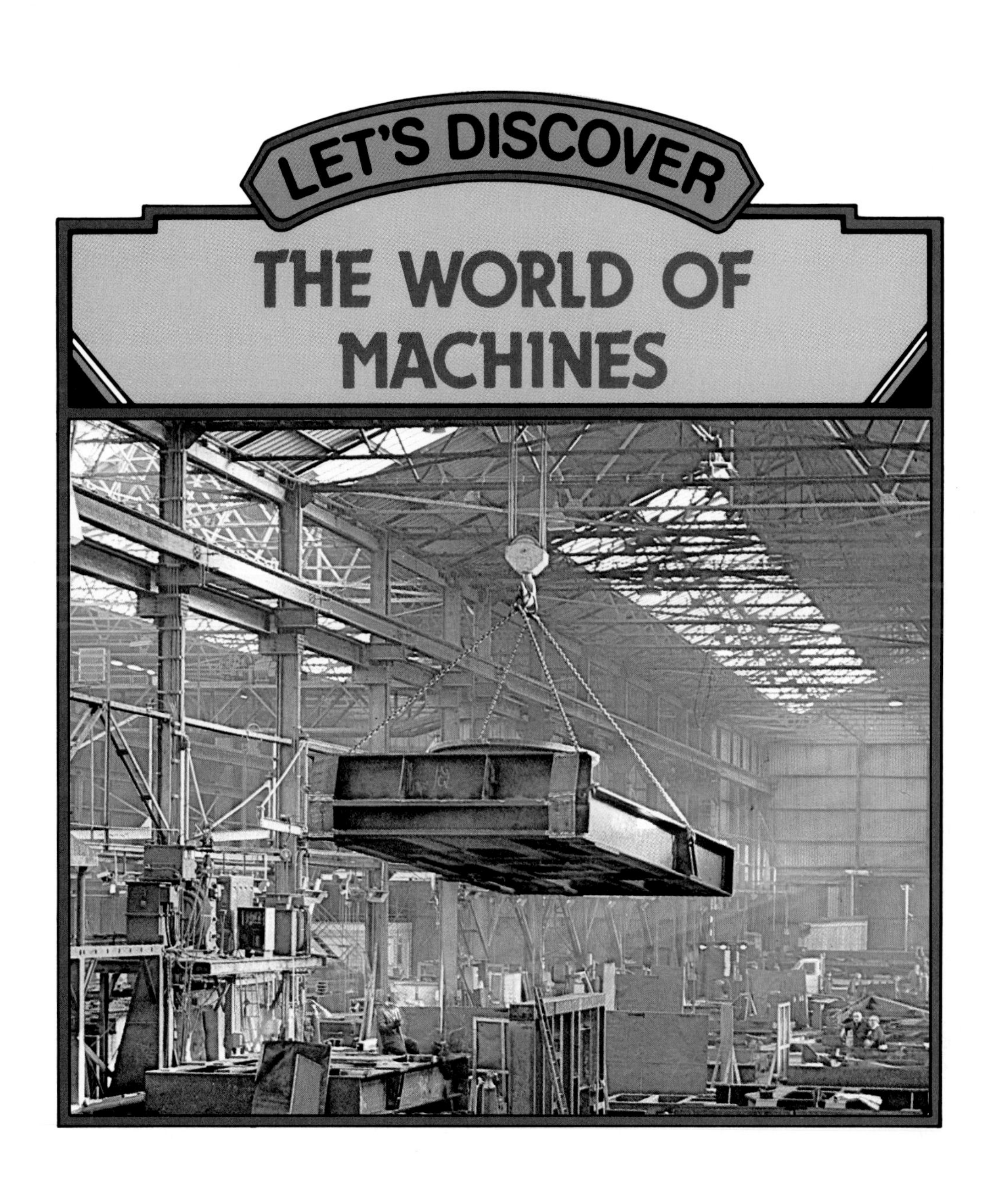

RAINTREE PUBLISHERS
Milwaukee • Toronto • Melbourne • London

Contents

MACHINES ALL AROUND

Long ago, people had no machines. They had to work by hand. Now, we have many machines. They help us in many ways. We use them at home, on farms, and in factories. People use machines on land, on the sea, and in the air.

brake
brake
gears
pedal
tires
road drill
wheels
brake

Wheels and gears

The wheel is a simple machine. It is also an important part of more complicated machines. The wheel was invented about 7,000 years ago. People wanted to carry heavy loads. To do this they put wheels on carts. The wheels could roll on rough ground.

Wheels are useful for the toys you build. Wheels make things roll along. You can use them to make pulleys. Pulleys can lift things.

Our life would be very different without wheels. There would be no cars, clocks, or windup toys. There would be no rides at the fairground. Wheels are important for engines and turning machines.

You can use round boxes and corrugated paper to make gear wheels. The slots on the edges fit into each other. The wheels then turn in opposite ways.

Before the wheel was invented, people dragged things along the ground. Then they found it was easier to move things on rollers. Later, they cut thin slices off the rollers. These made wheels.

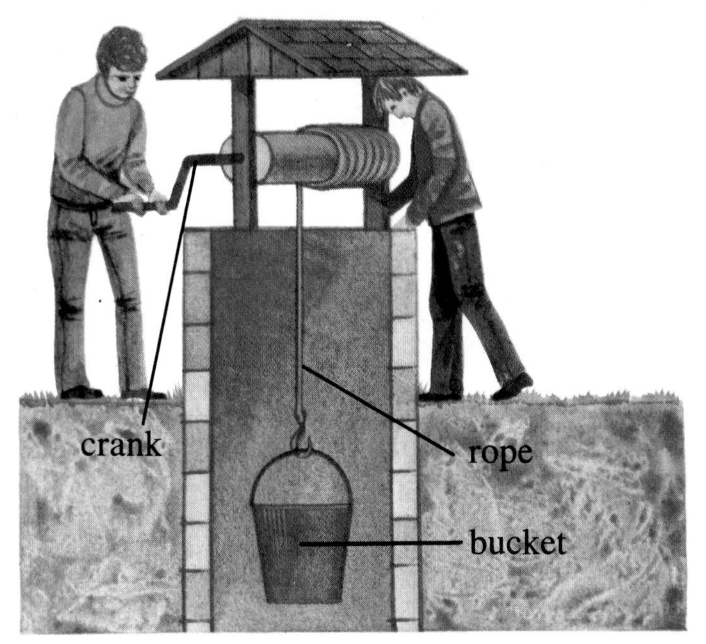

The roller above has a crank. It winds up the rope and bucket. This machine is called a wheel and axle.

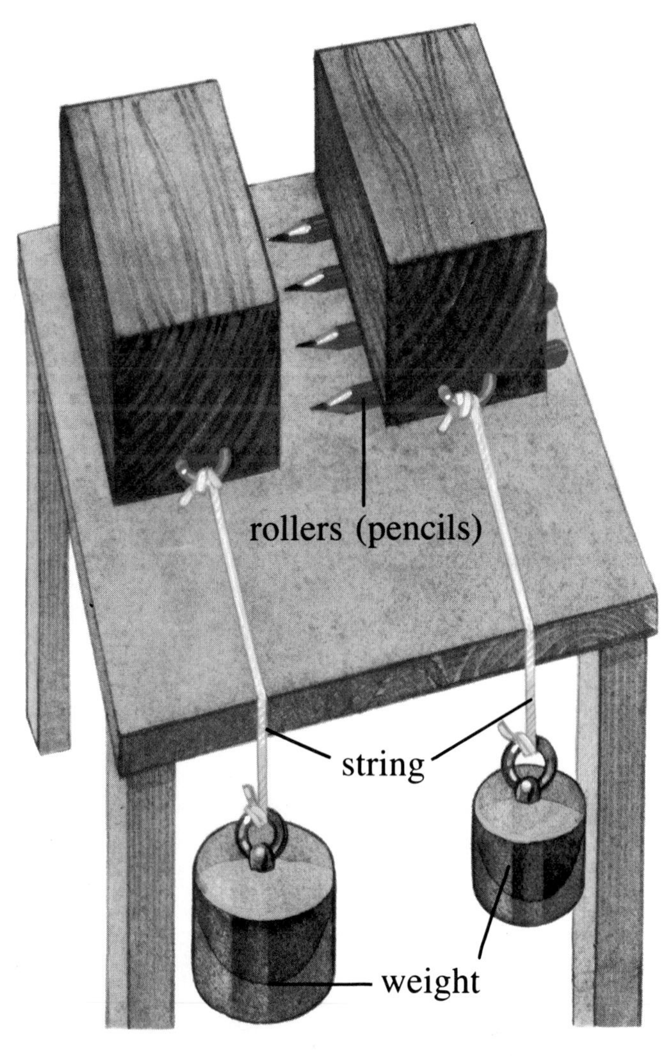

Find out if it is easier to roll something or to drag it. You need two objects that are alike. Put rollers under one of them. Which one moves more easily? Which needs a stronger pull?

The pulley is a kind of wheel. It has a groove all the way around it. A rope fits into this groove. The rope moves easily. Pulleys help us lift heavy loads.

Levers

The lever is the simplest machine. A bar is a lever when it is on a support. It moves on the thing that supports it.

A shovel is also a lever. The blade moves against the soil. The handle is longer than the blade. This helps move the soil.

A pair of nutcrackers is a lever. The long handles give us more power to crack the nut.

Look at the seesaw in the picture. The seesaw is a lever. It is supported in the middle. The plank rests on the stand.

A burglar uses a bar. He uses it to pry things open. The long handle gives extra power.

Wedges and screws

A wedge is a simple machine. Its shape is very important. If you lay a wedge flat, you will see that it slopes. A ramp is a wedge. People in wheelchairs roll up and down ramps. They find a ramp's slope helpful. Stairs are a kind of wedge. They make it easier to go upward.

A wedge can be used to split logs. The thin end is pushed farther and farther down into the wood. The slope pushes the sides of the log apart. The log splits open into two parts.

Have you ever come down a spiral slide? It is a slope that goes around in the shape of a spiral.

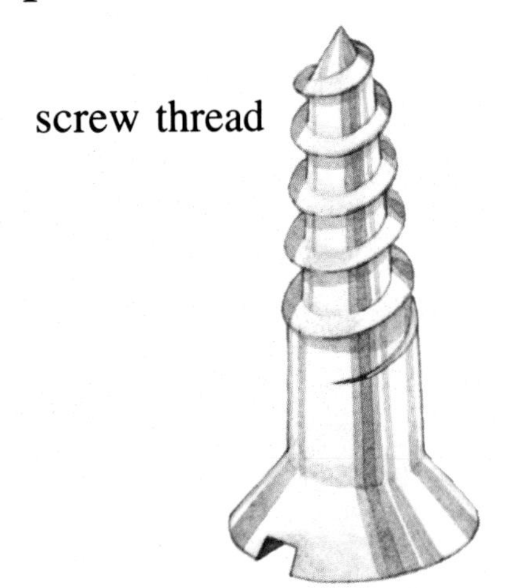

The spirals of a screw look like the slide in the picture. These spirals are called the thread.

IN THE HOME

People use many machines that help them in their houses. In the kitchen, they use machines for peeling, chopping, and grinding food. Mixers mix things for baking. Refrigerators help to keep food fresh. How many machines can you see here?

clock
hedge trimmer
lawn mower
radio
can opener
vacuum cleaner
washing machine
pepper grinder
toaster
iron
kettle
blender

Washing and cleaning

Washing clothes and wringing out the water by hand is hard work. The picture below shows a washing machine with two tubs. One tub washes the clothes. The other spins the clothes and flings the water out. The machine at the right does both jobs in just one tub.

It is hard to clean carpets by hand. The best way is to use a vacuum cleaner. It sucks up the dust. The picture at the left shows a cylinder cleaner. The one below shows the inside of an upright cleaner. The fan sucks up the dust. A bar beats the carpet to loosen the dirt.

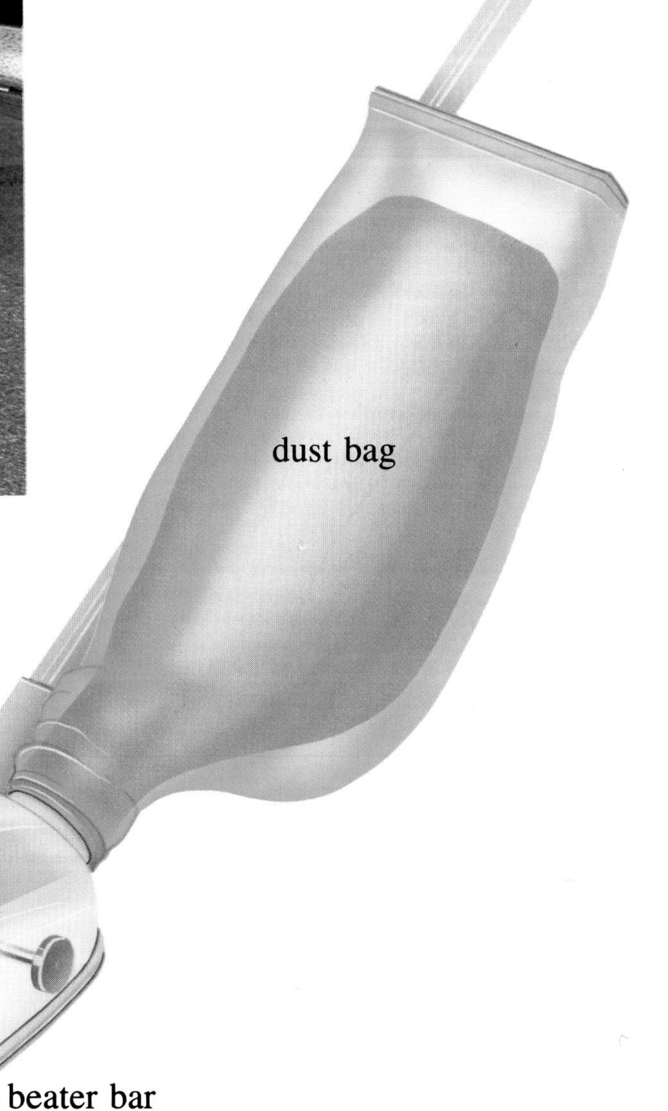

Tools and gadgets

We use many different tools and gadgets in the home. They are all machines. Most of them are made up of different parts. These machines often contain levers, wedges, screws, and wheels. Here are some of the many machines we use in our homes. They help us in many ways.

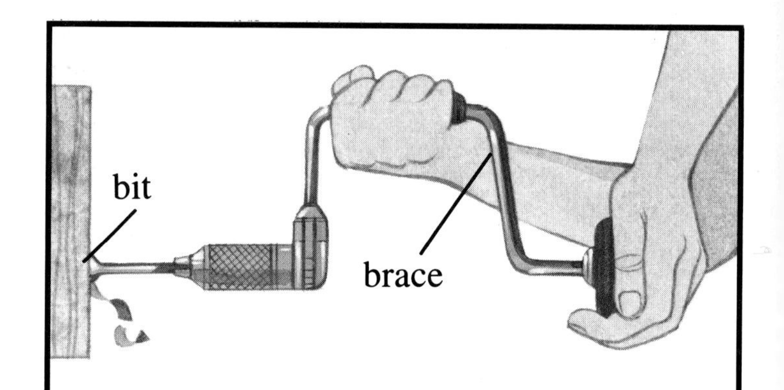

This brace and bit drills holes in wood. The brace turns the bit.

This can opener grips and turns the can with two wheels.

The drill in this picture uses gears to turn the bit.

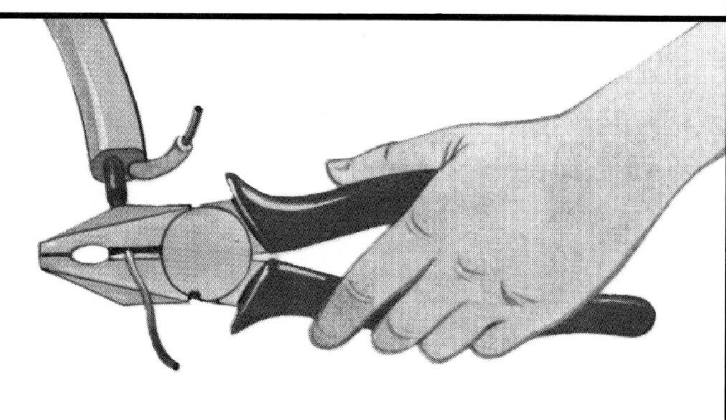

Pliers are used for gripping and cutting wires.

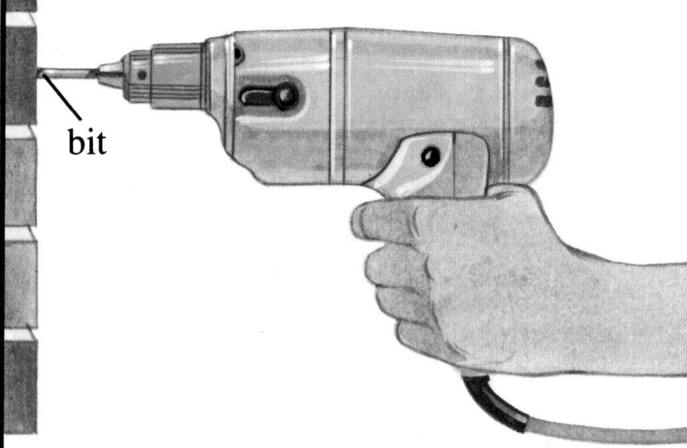

An electric drill uses an electric motor to turn the bit.

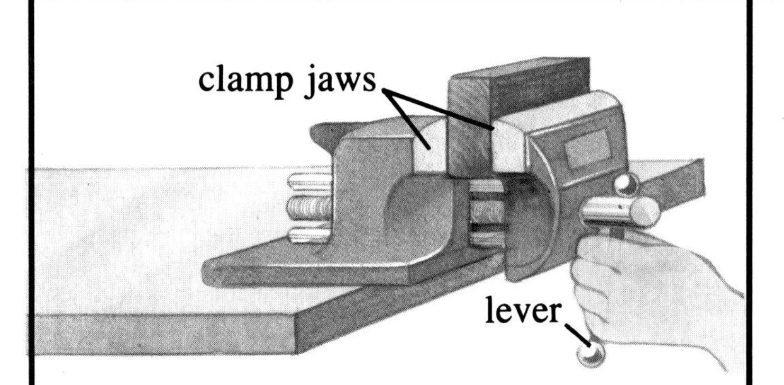

A vice is opened and closed by turning its screw with a lever.

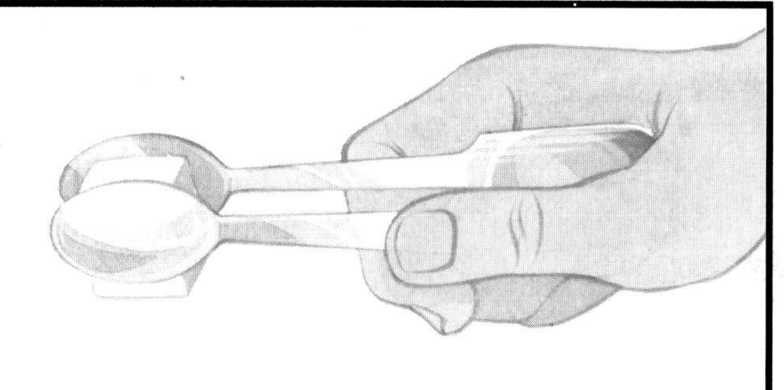

Sugar tongs are levers. We squeeze them to grip the sugar.

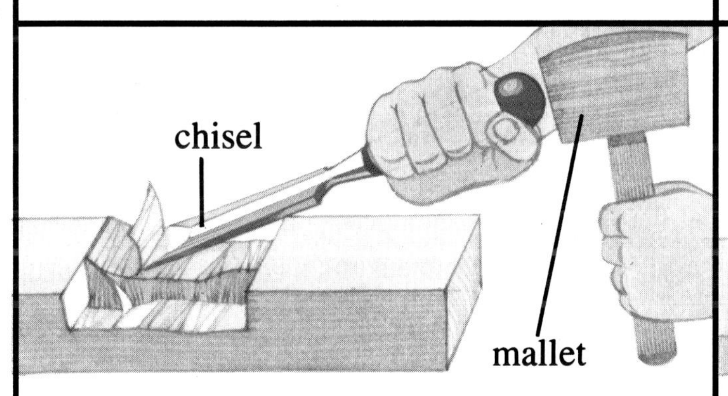

The cutting edge of a chisel is like a wedge. It splits the wood.

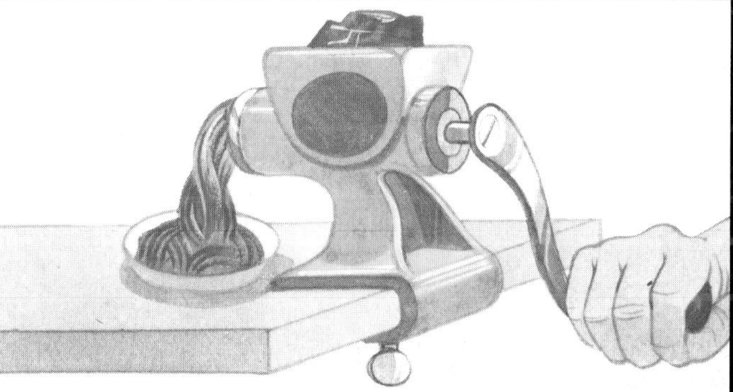

There is a large screw inside a grinder. It pushes the meat out.

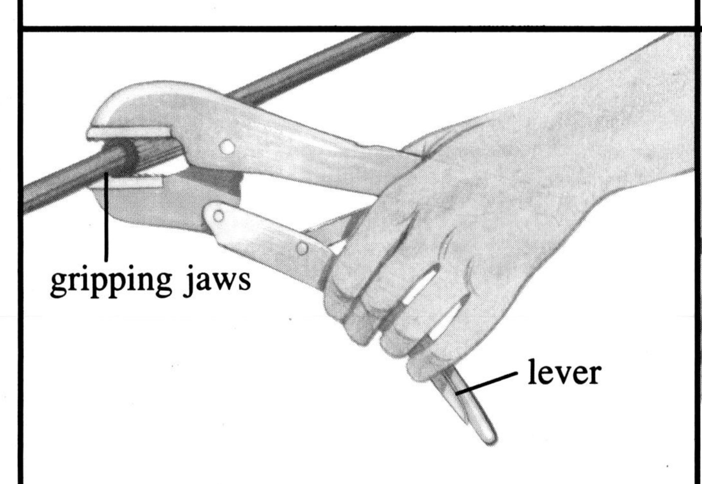

This wrench has lever handles. They make the ends grip tightly.

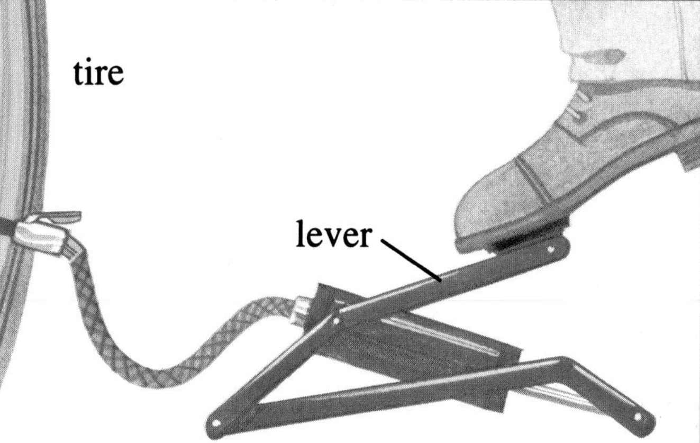

The lever on a foot pump is used to force air into tires.

Clocks and clockwork

The parts that make a clock go are also in other things. They are called clockwork. Clocks need something more to run steadily. They have a pendulum or a tiny wheel.

This watch at the right works by electricity. It has a battery.

The clocks below are driven by clockwork that is powered by hanging weights.

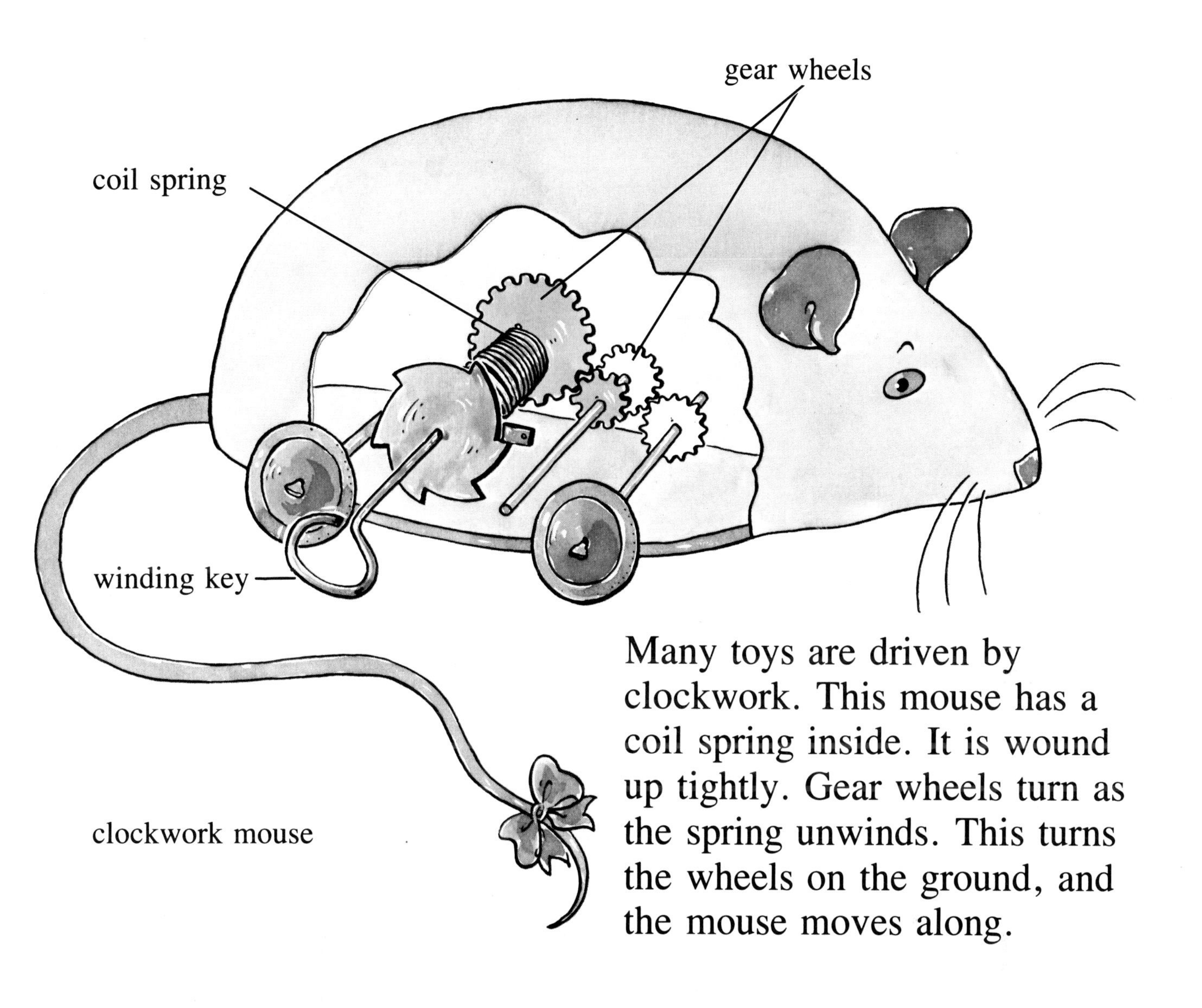

clockwork mouse

Many toys are driven by clockwork. This mouse has a coil spring inside. It is wound up tightly. Gear wheels turn as the spring unwinds. This turns the wheels on the ground, and the mouse moves along.

This is the inside of a music box. It runs by clockwork when its handle is turned. This turns a gear, which turns a cylinder. You can see tiny pins on the cylinder. These tiny pins hit pieces of metal called reeds. The sound of the pins hitting the metal makes a tune.

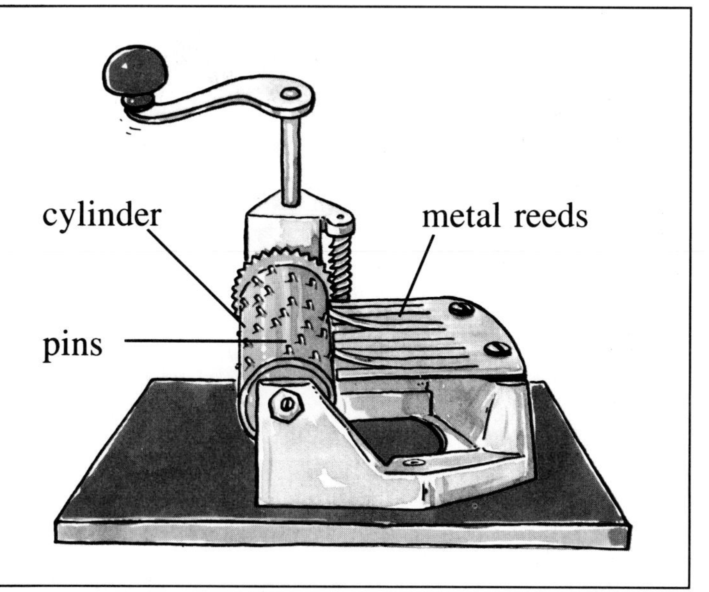

In the yard

A lot of different work needs to be done in a yard. The lawn must be cut. Hedges must be trimmed. Bushes need to be cut back. We use different machines to do all this work. Some are simple machines with few parts. Others, such as the lawn mower, are complicated machines.

An electric hedge trimmer has two blades with sharp teeth. They look like blades of a saw. One stays still. The other moves rapidly back and forth.

This electric lawn mower has a sharp blade that lies flat on the grass. The blade cuts as it goes around. The cord carries electricity from the house to the motor.

Pruning shears are very helpful. They are used for cutting thick stems. Only the top blade is sharp. These shears will not cut grass. Grass shears must have two sharp blades.

The mower below must be pushed. This turns its roller. The roller turns a large gear wheel. It turns smaller gear wheels, which turns the cutter. The grass is cut when the cutter pushes it against a fixed blade underneath.

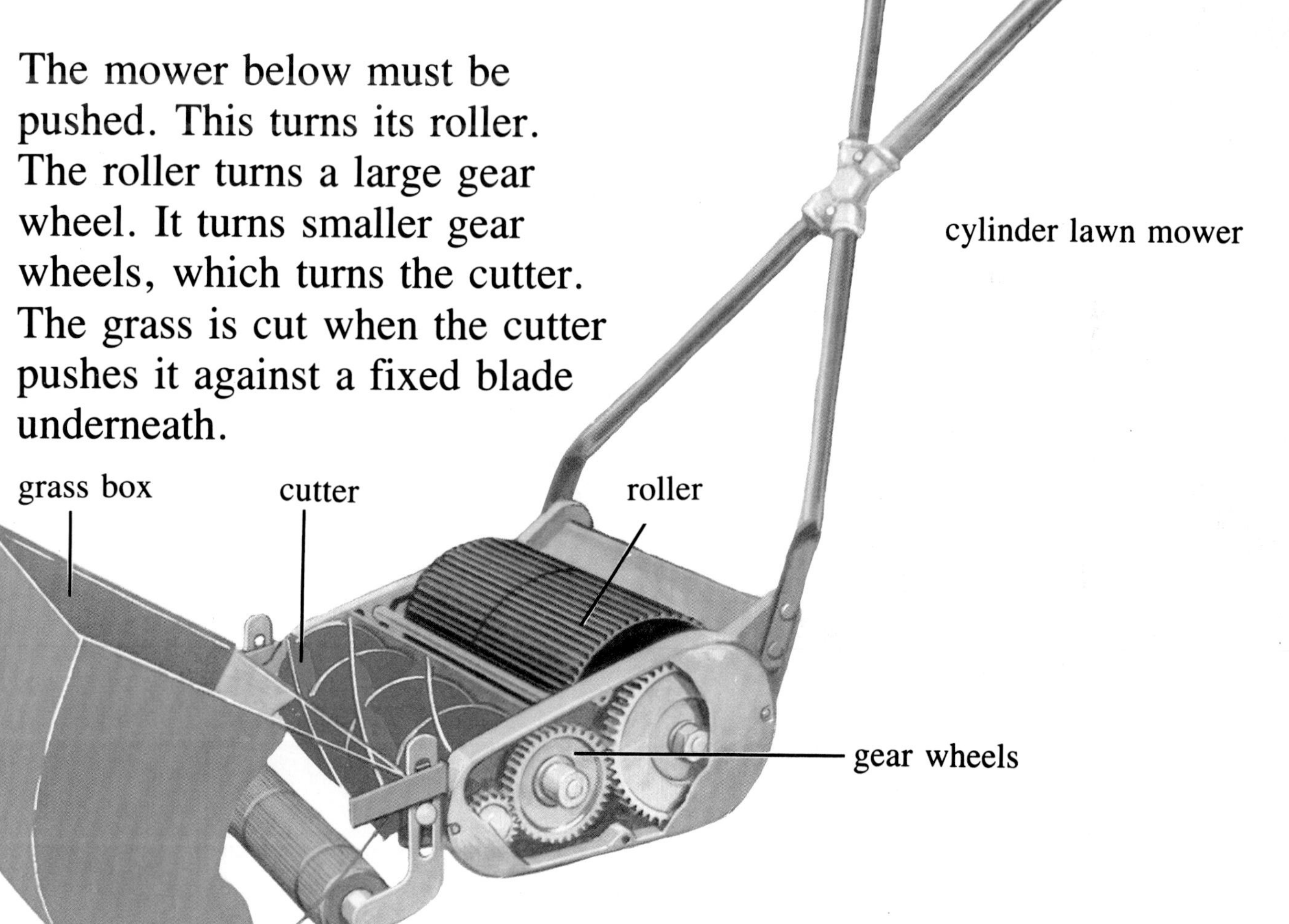

milking barn
milk container

ON THE FARM

Years ago there were more farmers than there are today. Now fewer farmers produce more food than ever before. Machines help them do this. Machines prepare the soil for planting. They plant and harvest the crops. The most useful farm machines are the tractor and combine harvester.

Farm machines

Today few horses work on farms. Tractors do their work instead. Machines can be used all day long. They pull plows, harrows, rollers, and drills. Drills sow seeds. The other machines prepare the soil. The tractor below is pulling a drill that plants seeds in an unplowed field.

The tractor above is pulling a plow. The blades of the plow dig deeply into the soil. They turn it over. Most tractors have large rubber tires. They give tractors great pulling power. Tractors work best on flat ground.

Harvesters

Farmers use machines to spray crops with chemicals. This helps keep the plants healthy. Machines also scatter fertilizer over the plants. This makes them grow well. When ripe, these crops are harvested mostly by machines.

The harvester above does two jobs. It cuts the crop. It also separates the grain from the stalk. It is a combine harvester.

Grass is cut by forage harvesters. They strip off grass to feed to animals in winter. Some grass is left to dry. It is called hay.

The harvesting machine above is a potato digger. It lifts the potatoes out of the ground and cleans them. A tractor pulls it.

This large machine is used to harvest corn. First it cuts the stalks. Then it takes the ears of corn off the stalks.

IN THE TOWN

Often old buildings are pulled down. Then new ones are put up. Heavy machines help to do this. They clear the ground. They dig out foundations. They pour concrete. They put up frames of new buildings. Many other machines will be used in this new office building.

truck crane
excavator
47
T. COX & SON

Cranes

There are many kinds of cranes. They are used for lifting. Each crane has a long arm. This arm can usually swing in all directions. The arm carries a long cable with a hook. The cable runs through pulleys. It has a heavy block at one end, called a block and tackle.

Have you ever seen a truck crane? It is moved on its truck from place to place. Part of it rests on the ground when it is working.

The picture at the right shows an overhead crane. It can lift and carry objects sideways. It moves all along the building.

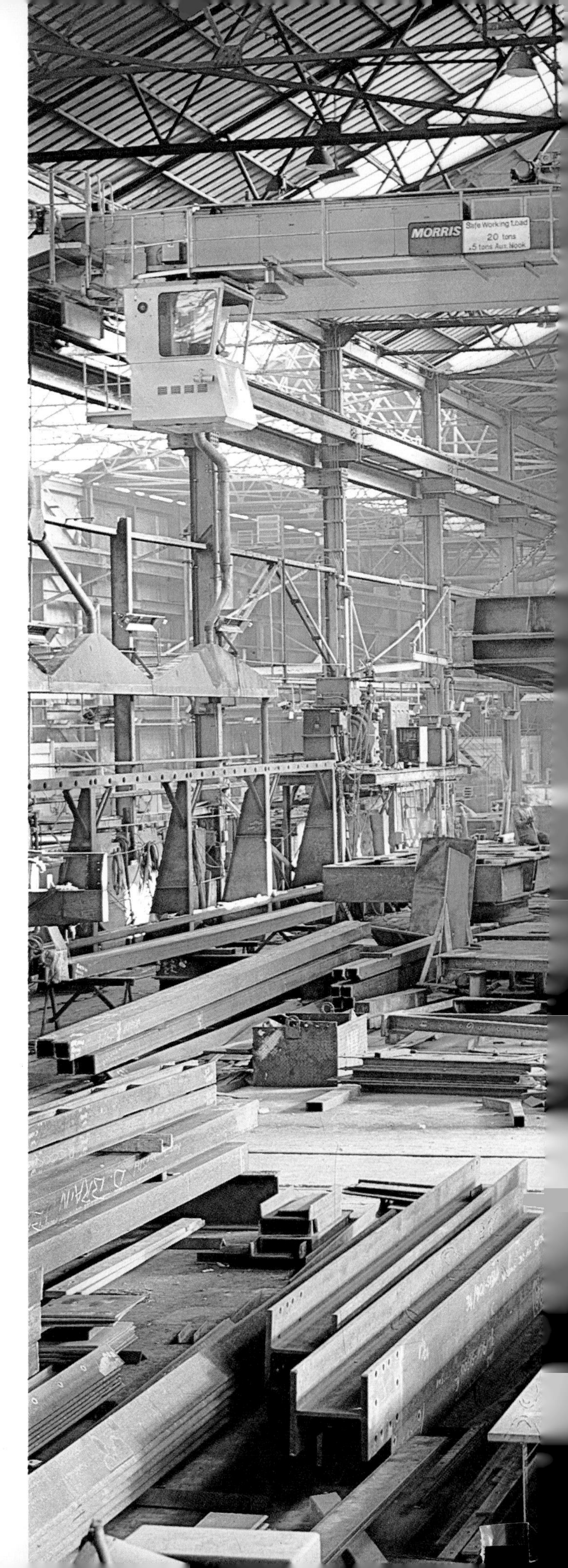

A tower crane is used for putting up tall buildings. The tower is made taller as the building gets higher. The long arm moves around and over the building. A trolley travels along the arm. Pulleys and ropes hang from the trolley. This crane can carry things over a very wide space.

Digging machines

Digging with a pick and shovel is a very slow job. It is hard work. Drills and digging machines now make this work fast and easy. Digging machines are called excavators. They look like cranes, but they have a scoop instead of a block and tackle.

Excavators can do other jobs. This one is knocking down a wall. The arm and bucket do the work.

When you see a road being fixed, look for large noisy drills. They work by air pressure. Air is pumped into the drill. A part called a piston moves up and down. This makes the end of the drill shake back and forth. It hammers into the road. That is what makes it such a noisy machine.

Below you can see a grab excavator. The scoop clamps together to grab a load.

Elevators and escalators

Elisha Otis, an American, invented the elevator more than 100 years ago. Before that there were no very tall buildings. But elevators can carry only a few people at a time. A moving staircase or escalator can carry many more people.

The picture above shows an escalator on the outside of a building in Paris, France.

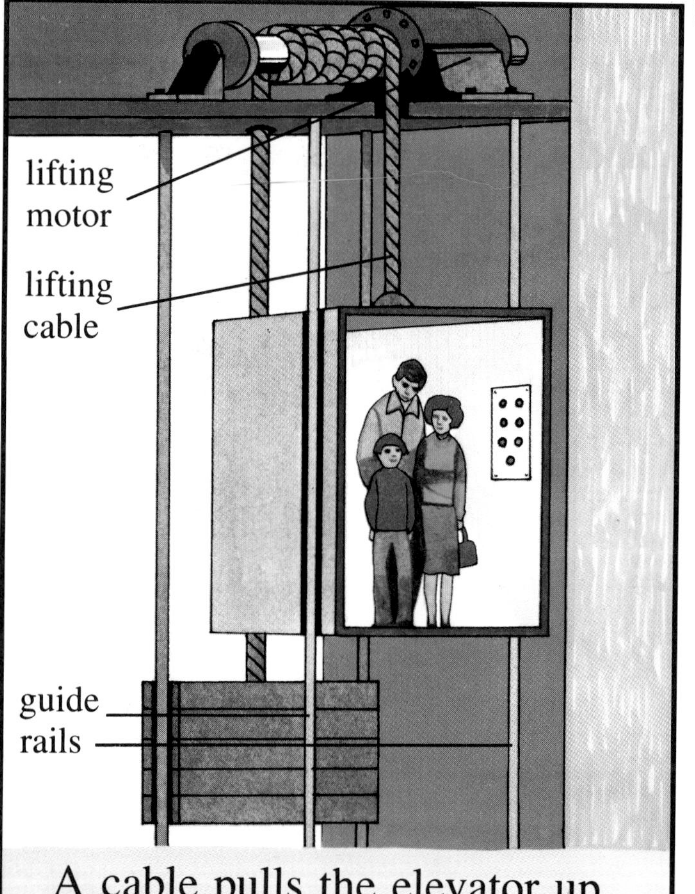

A cable pulls the elevator up and lets it down. An electric motor controls the cable. The elevator slides between guide rails.

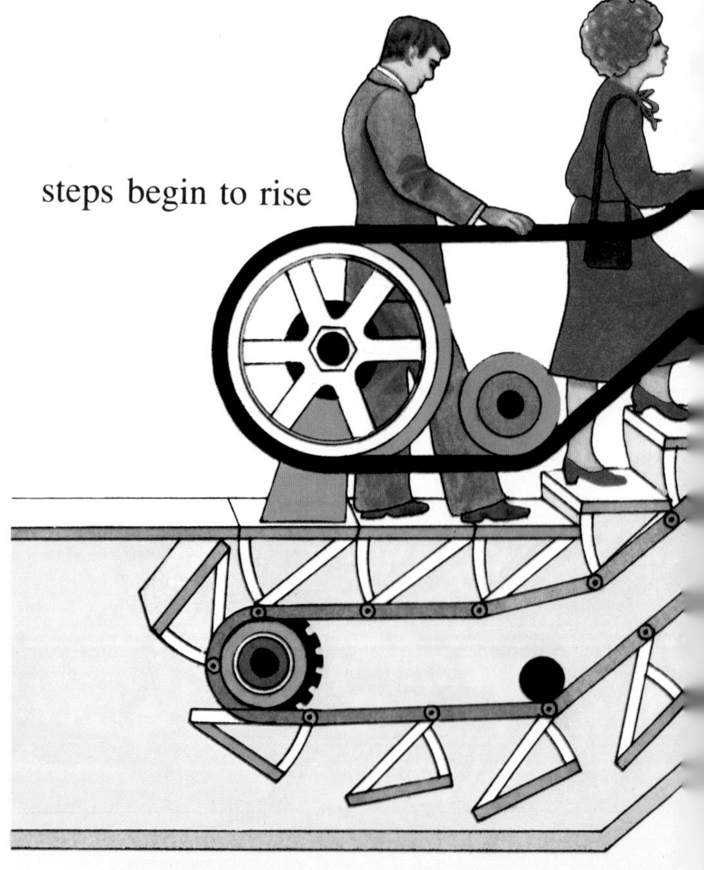

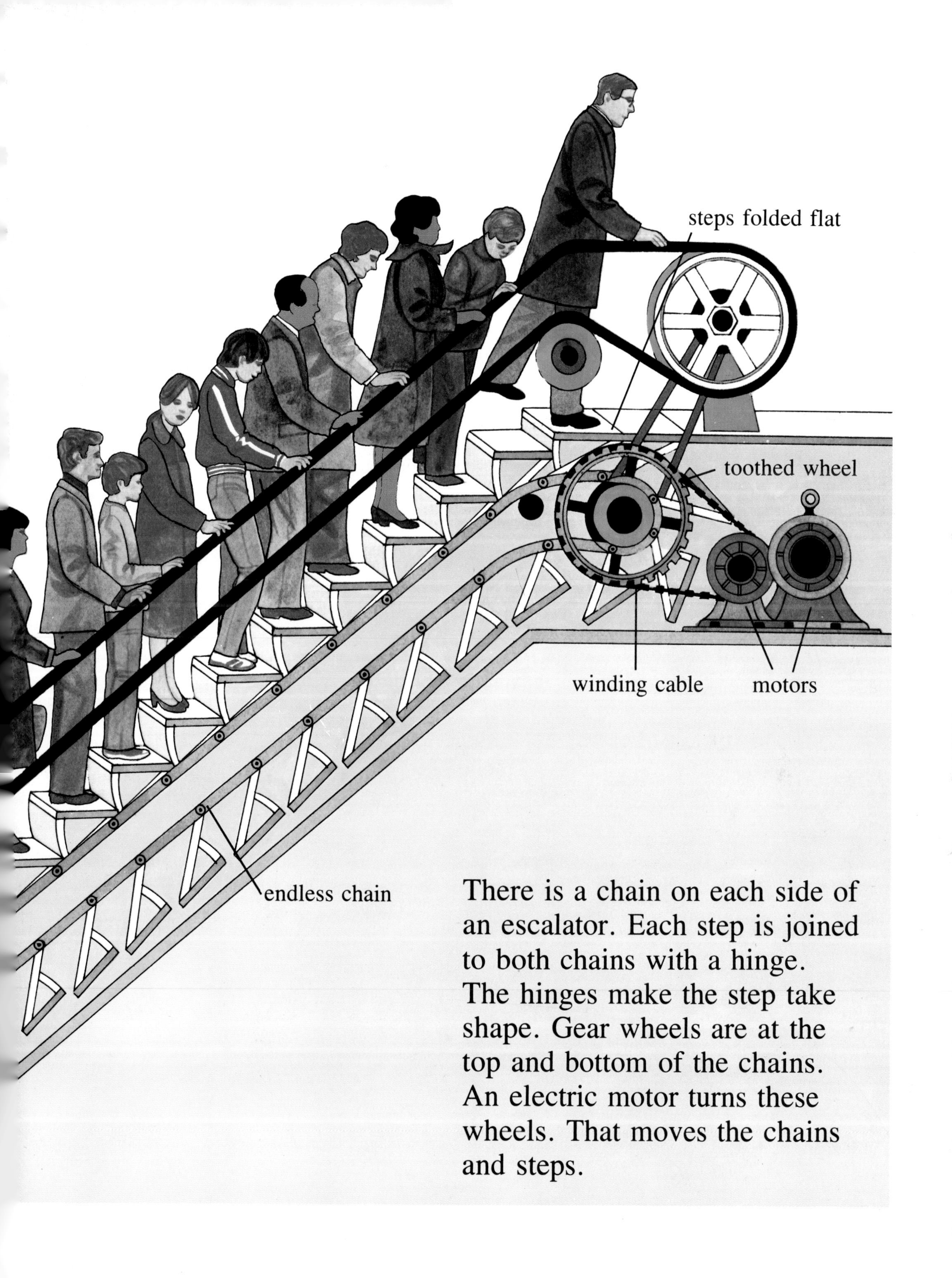

There is a chain on each side of an escalator. Each step is joined to both chains with a hinge. The hinges make the step take shape. Gear wheels are at the top and bottom of the chains. An electric motor turns these wheels. That moves the chains and steps.

Office machines

The typewriter was invented over 100 years ago. Early typewriters were big. Modern machines are much smaller. Many offices have photocopying machines. These make copies of letters and similar things. A Telex machine can send typed messages all over the world.

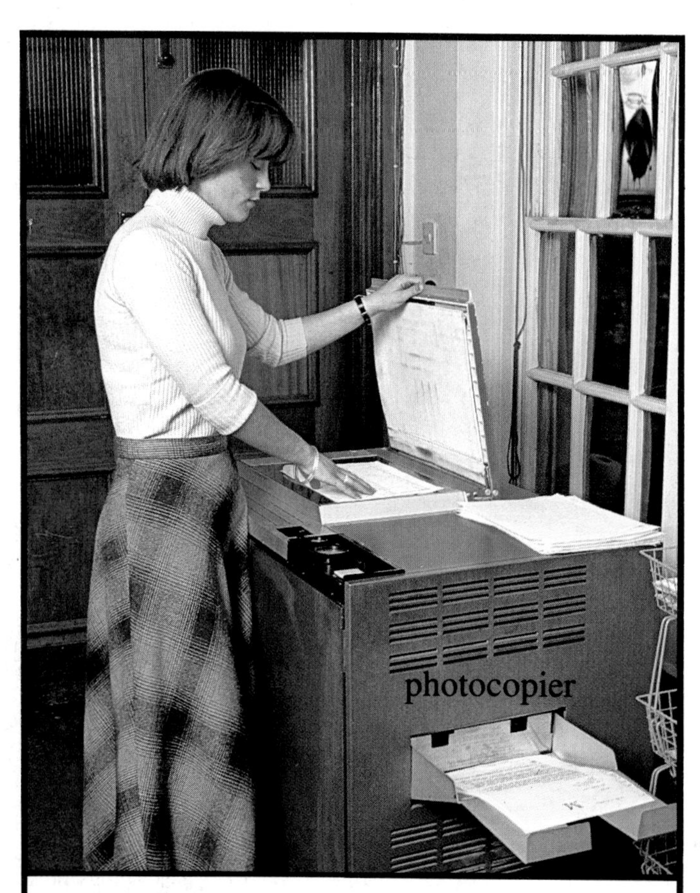

This woman is using a photocopying machine. It is powered by electricity. A letter is put on top of the machine. A copy of it comes out at the side.

The typist's fingers hit the alphabet letters through an inked ribbon onto a paper.

A part of this typewriter looks like a golf ball. It moves very quickly up, down, and around. It stamps letters onto the paper.

typewriter

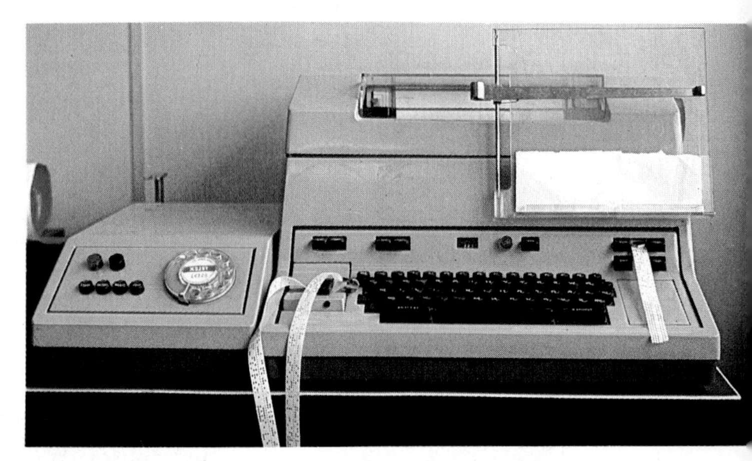

This Telex machine changes typed messages into electrical signals. The signals travel along telephone wires. A machine at the other end, called a teleprinter, then types out the messages.

IN THE FACTORY

Many years ago factories were small. The workers used their hands to make and fix things. The work was slow. Modern factories are large. Their workers use many kinds of machines to speed up work. These people are working in a furniture factory.

The man below is using a power saw. He is shaping planks of wood. They are for a chair.

The wooden parts are put together. This makes the frame. The man above is fitting wire springs into the back and seat.

All day long workers are busy in other parts of the factory. This machine is cutting material to make the chair covers.

Here the material patterns are being sewn together. Each chair cover is made the same size. It will fit any chair of similar size and shape. This work goes very fast.

The man at the left is putting plastic foam on a chair frame. This makes the chair comfortable to sit in. Finally, he will put the cover over the foam. You can see the finished chair above. It is ready to be sold.

Conveyors

In many factories, things must be moved quickly from place to place. A conveyor belt can be used to do this. The belt is made of firm material. It moves on rollers. Sometimes, an overhead conveyor belt may be needed. This is a moving chain. Things are hung on the chain's hooks.

Some airport buildings cover a large area. Conveyors are used to carry people and luggage. It is a quick and easy way to move.

These workers are in a factory. They take the parts they need from a belt as it passes. The picture at the left shows a conveyor used in mining. This belt is shaped like a trough to hold loose materials.

Robots

In many factories, jobs are done by machines alone. These machines are called robots. They have arms and claws like the ones shown at the right. Robots can move these parts the way human arms and hands move. Would you like to own a home robot machine like the one below?

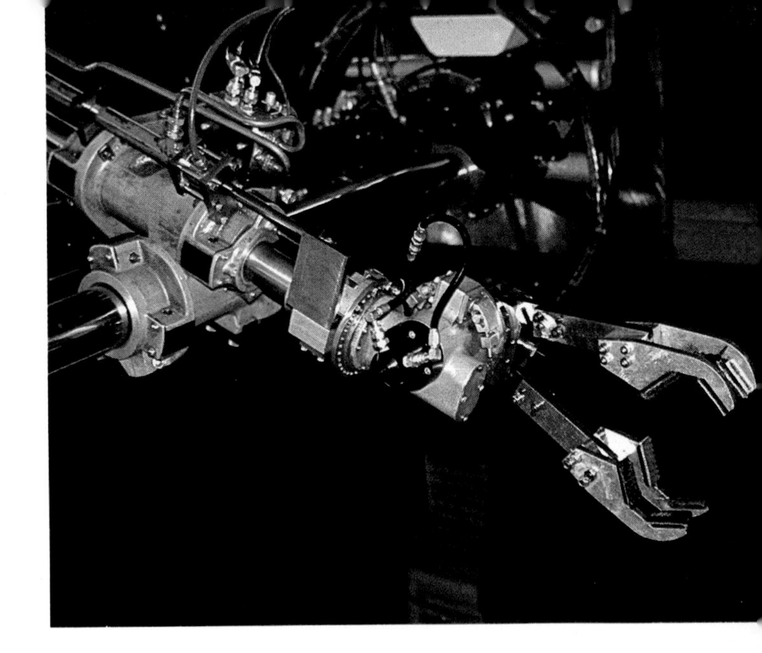

Machine tools

Many kinds of things are made with power tools. Each tool has an electric motor. People who work in metals use machine tools.

Some machine tools shape, stamp, and drill the metal. Others grind and polish it.

This man is using a machine tool called a shaper. It cuts thin slices off metal parts. The part he is shaping is held firmly. The cutting tool moves across it. This worker wears glasses to protect his eyes.

The pole lathe below was used long ago. String was wrapped around the part being shaped. The part turned when the string was moved up and down.

Here is a modern lathe. A wooden bowl is turned by a motor. A sharp tool is held against the bowl to shape it.

Making a wineglass

Here you can see how a wineglass is made. Sand, soda, and limestone are heated in a furnace. They get so hot they melt into a liquid. The liquid cools and turns into glass. The flat glass used in windows is shaped by a machine. Wineglasses are made mostly by hand.

1

First, a bubble is blown from melted glass. The blower uses a tube called a blowing iron.

2

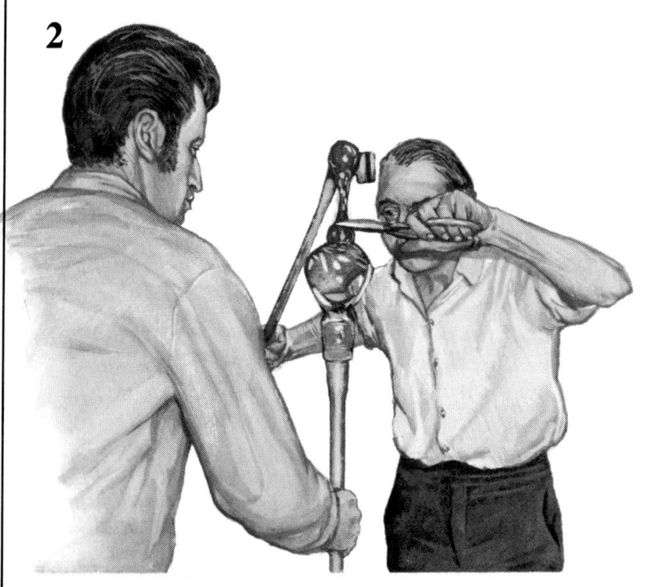

The bubble will be the bowl of the wineglass. Next, a piece of hot glass is joined to the bubble. It makes the foot and stem.

3

Now the stem of the wineglass is shaped. The hot glass is gently squeezed between two pieces of wood to form the foot. The wineglass is then broken off the blowing iron. It is heated and then slowly cooled.

4

The wineglasses are looked at carefully. The top of each perfect glass is cut off. The rough edge is smoothed.

5

The wineglass is now ready to be decorated. A pattern is painted on it for the cutter.

6

The cutter uses a grinding wheel. He cuts into the glass along the pattern lines.

Finally, the wineglass is dipped in a liquid to make it sparkle. It can now be used.

7

ON THE ROAD

Bicycles

Cycling is fun. It is a cheap sport because no fuel is used. Racing bicycles can travel as fast as 30 kilometers (18.6 miles) an hour. Mopeds are bicycles with engines. You can see one in the small picture. It has pedals. The rider uses them to start and to go uphill.

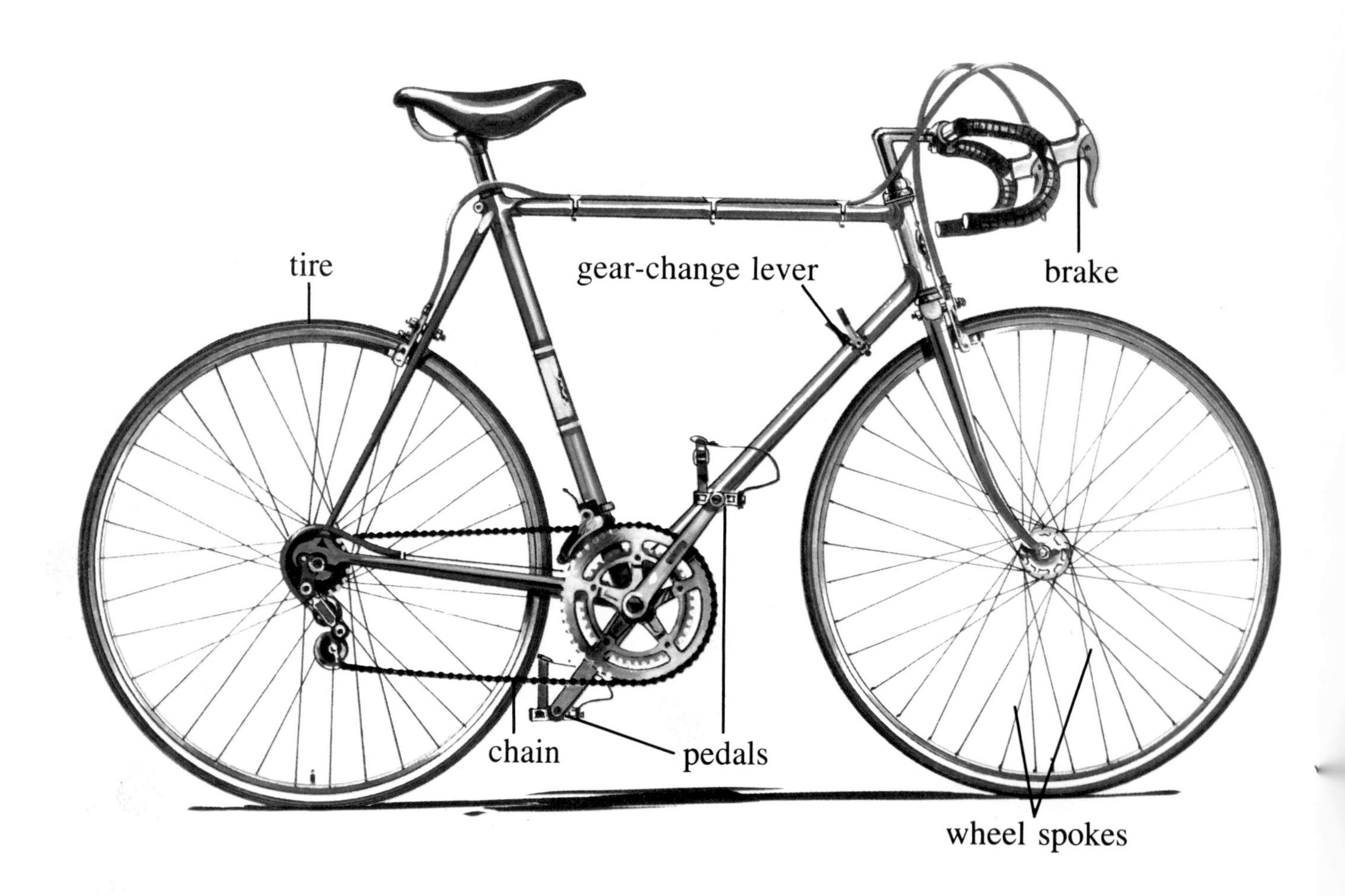

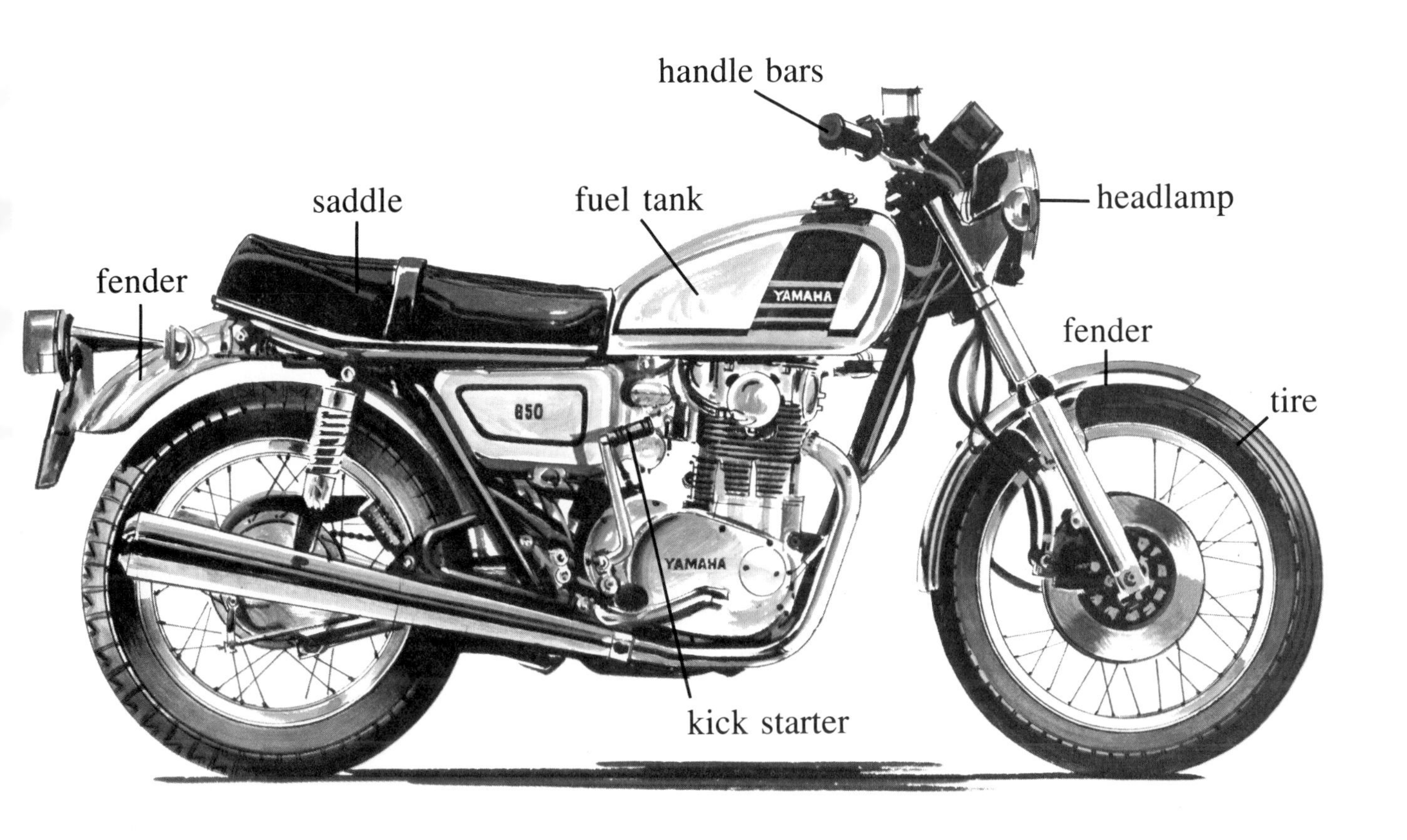

Motorcycles

Some motorcycles can travel at speeds more than 160 kilometers (99 miles) an hour. The rear wheel is usually driven by a chain from the engine. On some motorcycles, the engine is started by pushing on the kick starter. On others, it is started by an electric motor.

Inside the car

There are thousands of parts in a car. They work together in groups. The groups have been colored in this picture. The engine and parts that help it are blue. The electrical parts are red. Braking and steering parts are green, as are parts that help give the car a smooth ride.

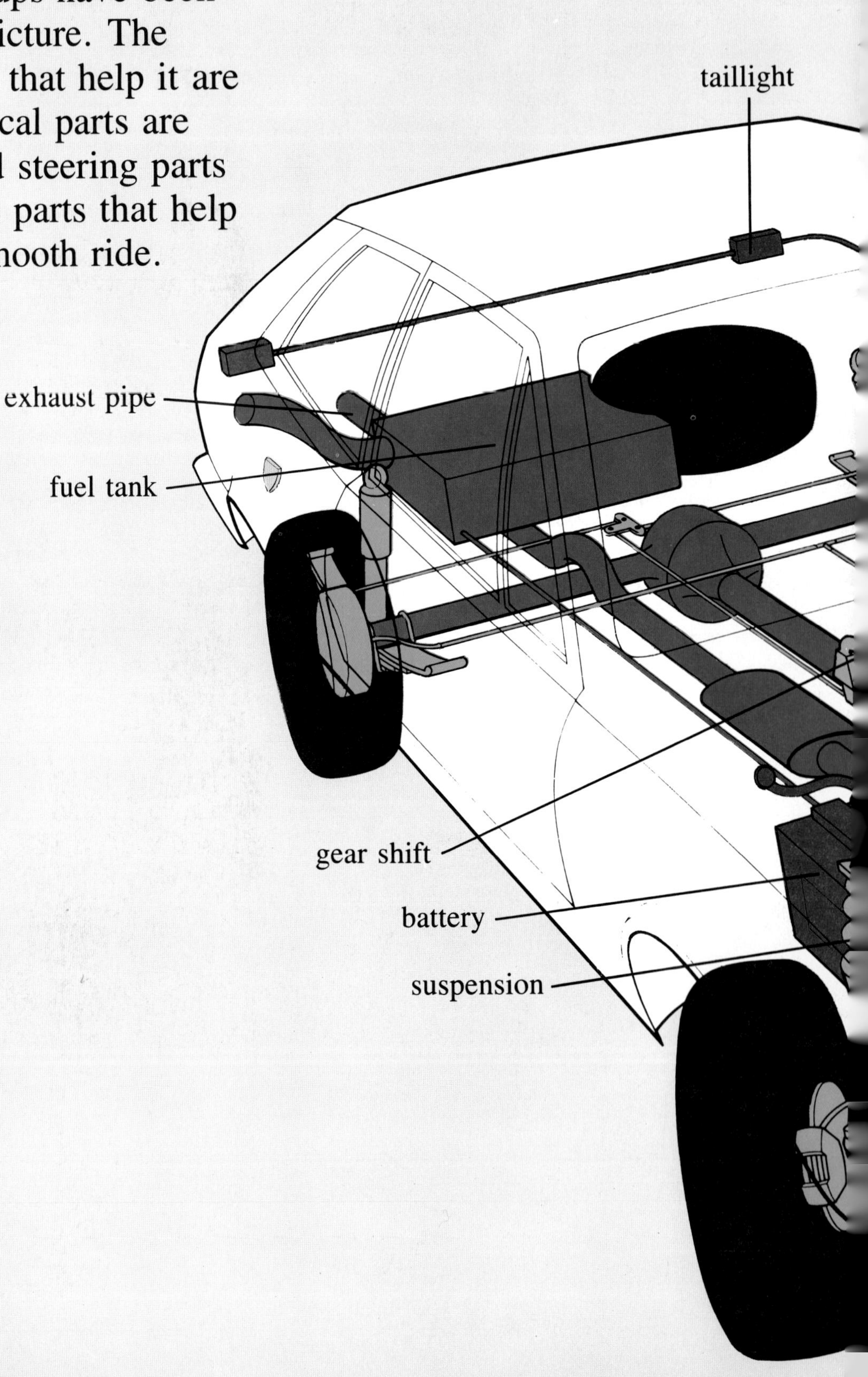

The engine can be at either end of a car. It can drive the front or back pair of wheels. The ordinary family car has a closed-in body. It is called a sedan. Cars called convertibles have a top that can be opened. Station wagons have a long body. They can hold many things.

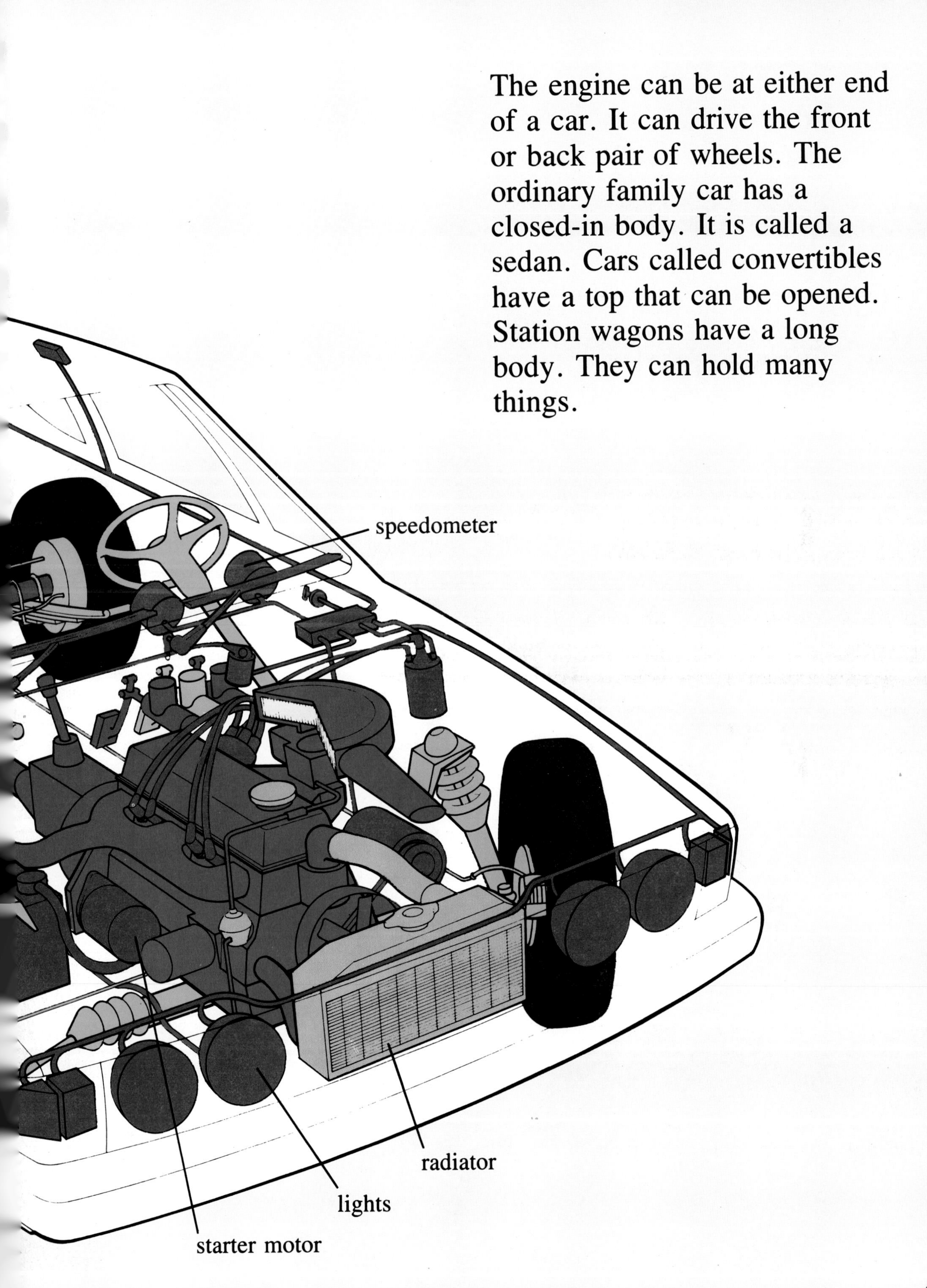

Parts of a car

Look at the parts that help the engine power do its work. As the car moves along, air cools water in the radiator. This water cools the engine. Transmission gears help the engine work at different speeds. Power travels along the drive shaft to the rear axle. This turns the wheels.

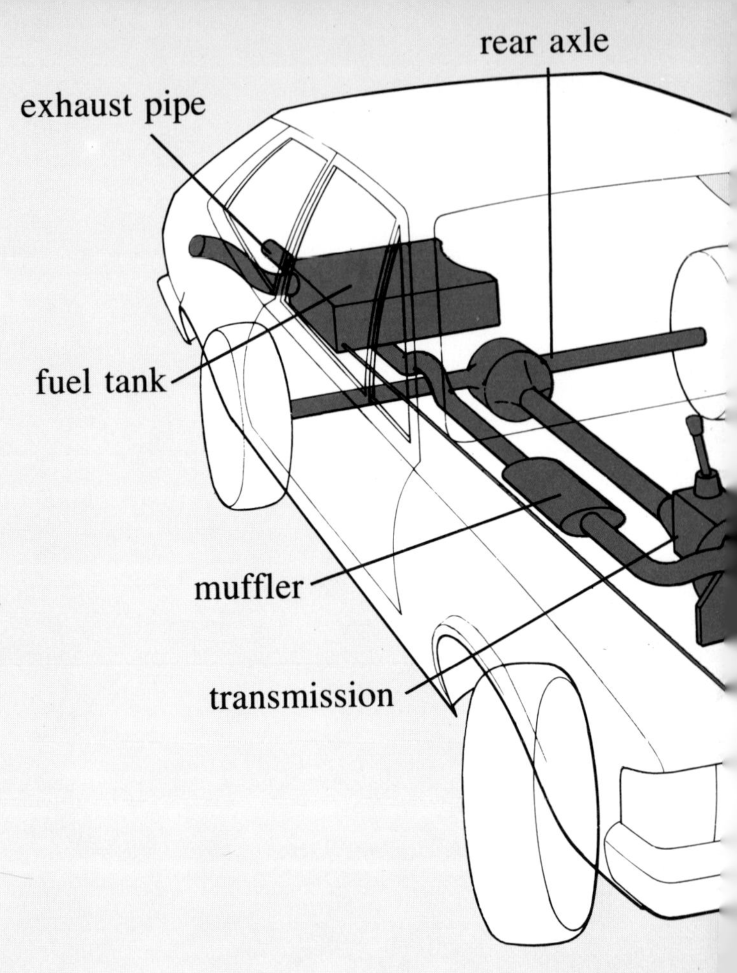

Braking, Steering, and Suspension System

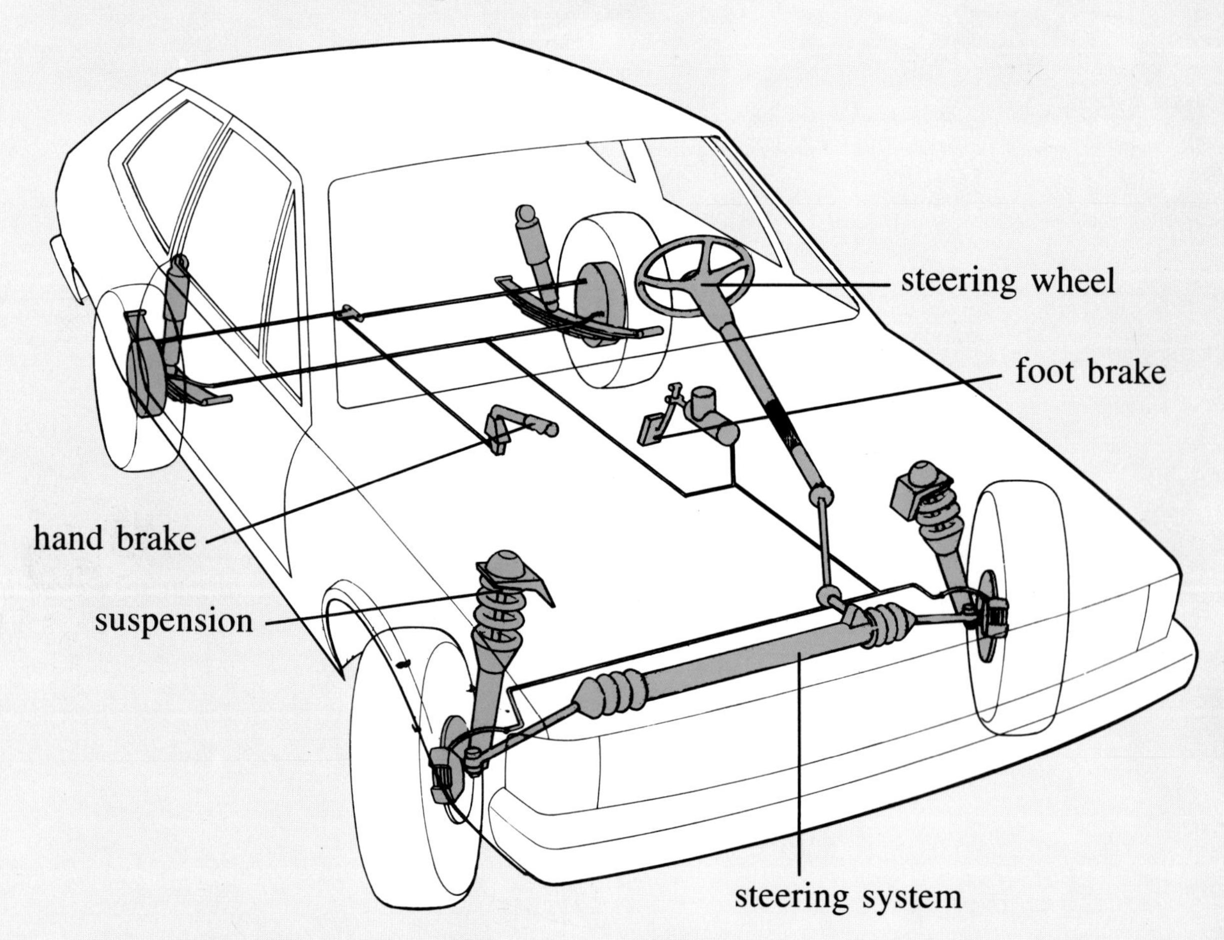

Engine and Transmission System

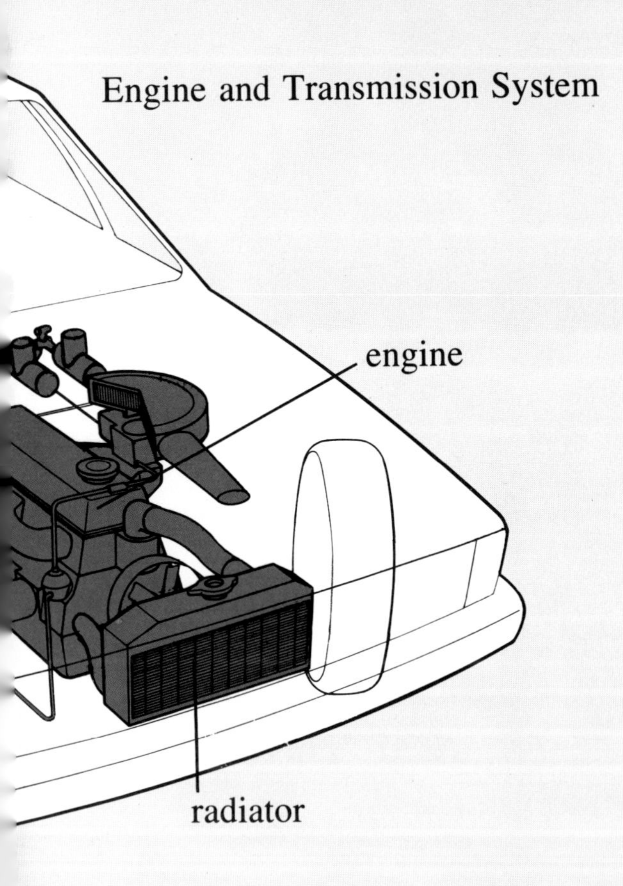

You can see the car's battery at the bottom of the page. The battery makes electricity. This gives power to start the engine. It also makes the lights and instruments work. The left picture shows the springs in green above the wheels. They help give us a smooth ride.

Electrical System

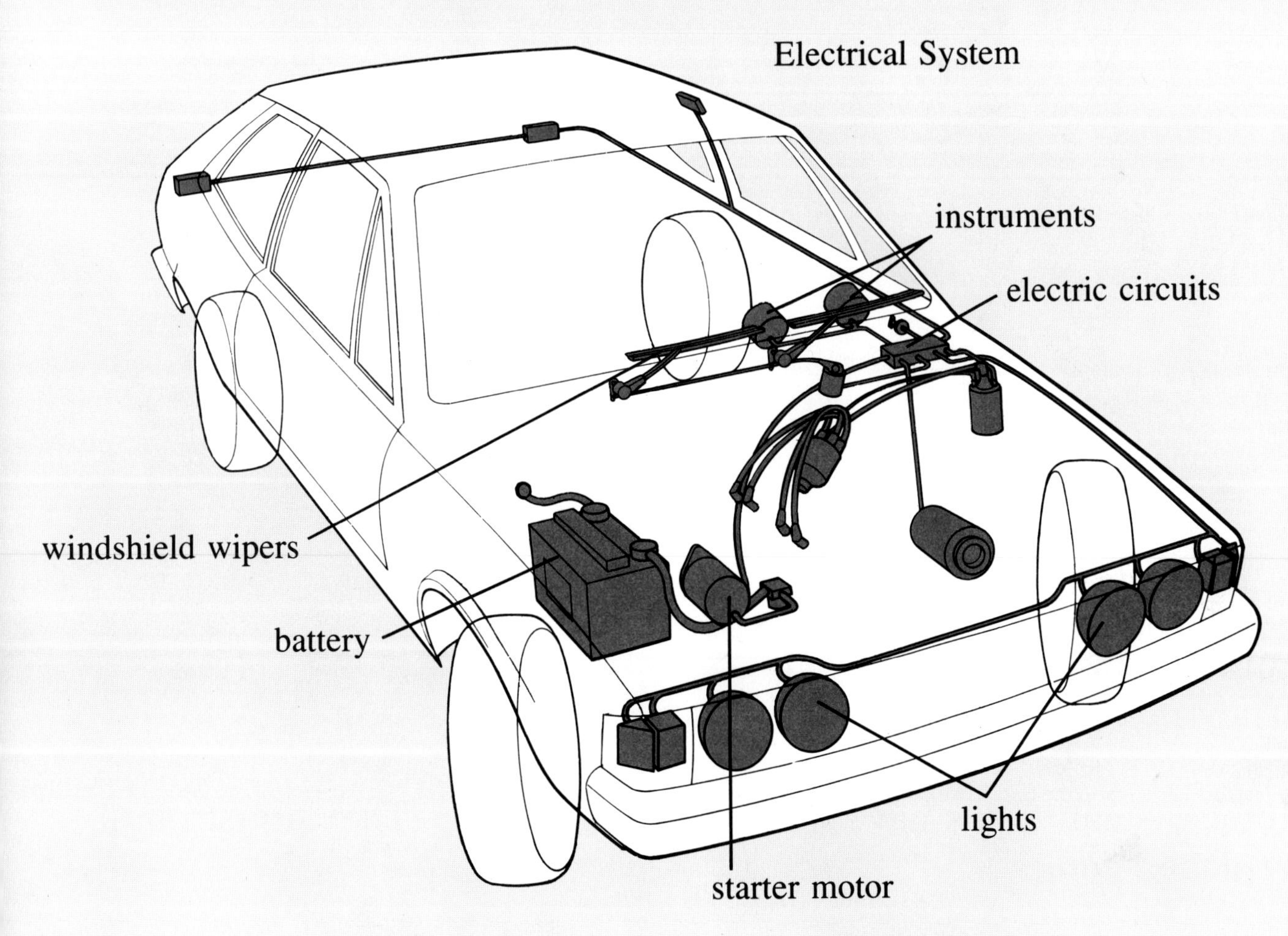

POWER

Animal power

Years ago there were no engines to drive machines. Muscle power was used instead. Many people used strong animals such as horses and oxen to do heavy work. In many parts of the world animals are still being used.

Years ago many factories used squirrels to power small machines. As the squirrel ran, the wheel turned. This drove the machine.

Ponies are used in some coal mines. They haul coal wagons. They pull the wagons along railway tracks. In most mines, small trains have replaced ponies.

Eskimos use dog sleds to travel across frozen snow. The dogs are huskies. They are strong, hardworking dogs.

In some countries oxen are used to run water pumps. The ox moves around. This turns gears. The gears move a big wheel with jugs that dip up water.

In some places the land is rough or very steep. There are no roads or railways. Donkeys do the hauling instead of machines. They are surefooted.

For hundreds of years humans turned wheels like this one. Such wheels powered cranes, pumped water, and ground grain.

Wind and water power

Windmills work by wind power. The wind turns their sails.

Years ago mills were used to grind grain into flour. Today some windmills are used to pump water from low-lying land. Water wheels are used to bring water to dry lands.

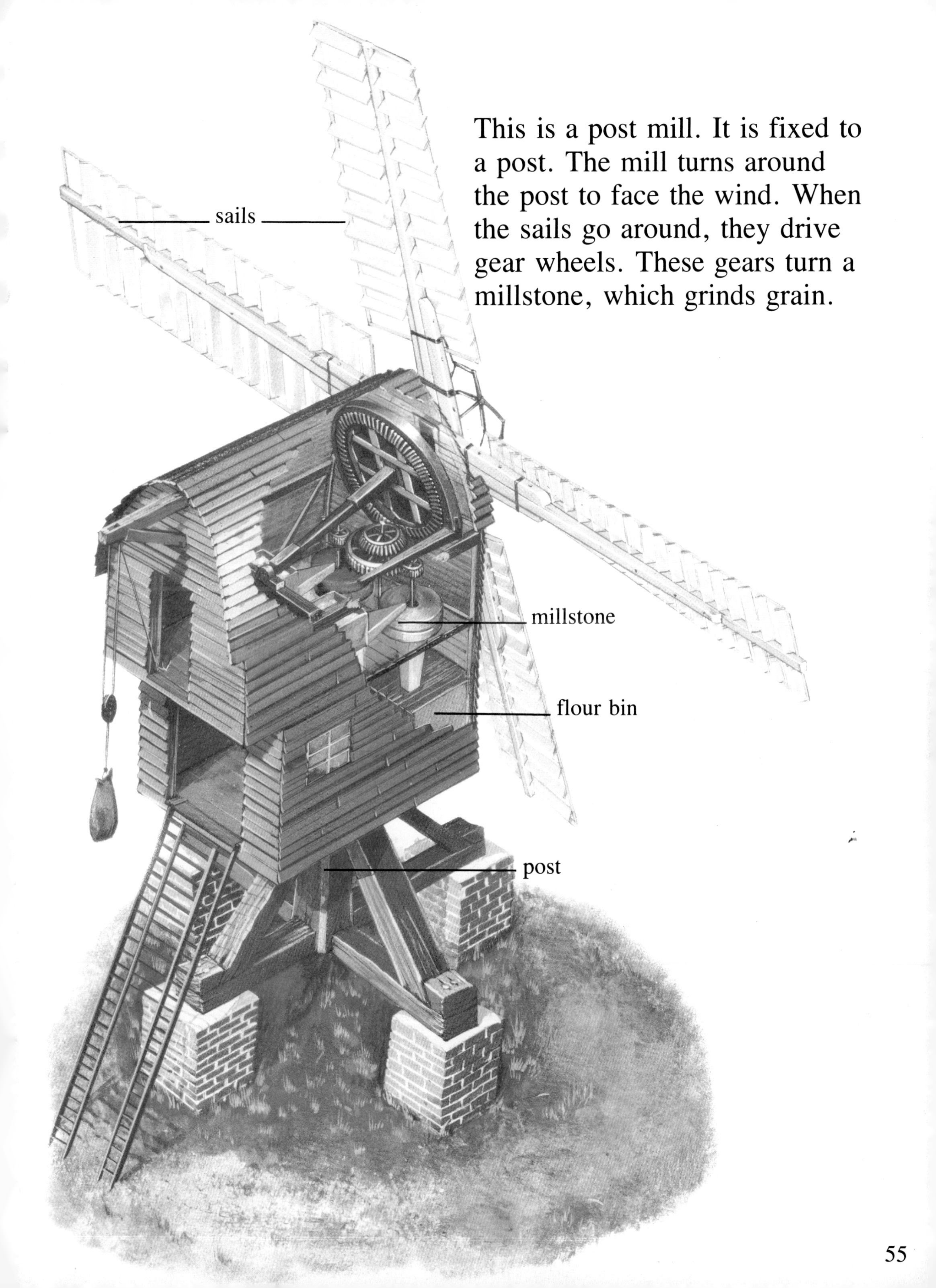

This is a post mill. It is fixed to a post. The mill turns around the post to face the wind. When the sails go around, they drive gear wheels. These gears turn a millstone, which grinds grain.

Steam power

Boiling water makes steam. This steam is full of power. We can use this power to drive engines. The steam pushes a piston into a cylinder. This action turns wheels or other engine parts. Railway travel started with a steam engine designed by George Stephenson.

This is a steam-powered fire pump. It was used before the first fire engine was made. Steam drove a pump that pumped water through the big hose.

This is a model steam engine. Steam is very powerful. There is enough power to pull the girl.

Look how this steam train works. The coal fire makes flames and hot gases. They pass through tubes and heat water in the boiler. The water turns to steam. Steam passes into a cylinder. There it pushes on a piston. The piston moves back and forth to turn the engine's big wheels.

Gasoline and diesel power

This is a picture of a car engine. It is made of metal. It is very heavy. It works by burning a mixture of gasoline and air inside its cylinders. Four movements take place in a cylinder. These are repeated many times. They make the engine work.

These pictures show what happens inside a cylinder.

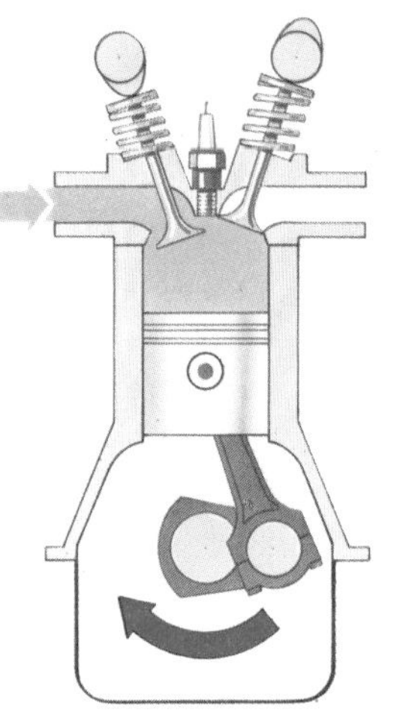

1. The piston goes down; fuel goes in.

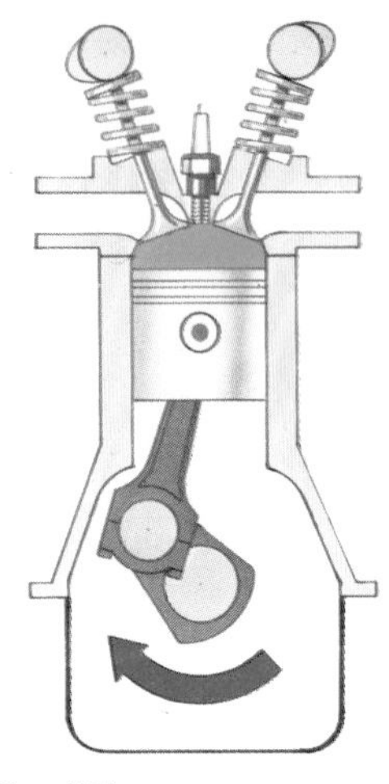

2. Piston goes up. It squeezes the mixture.

3. Mixture explodes. Piston is forced down.

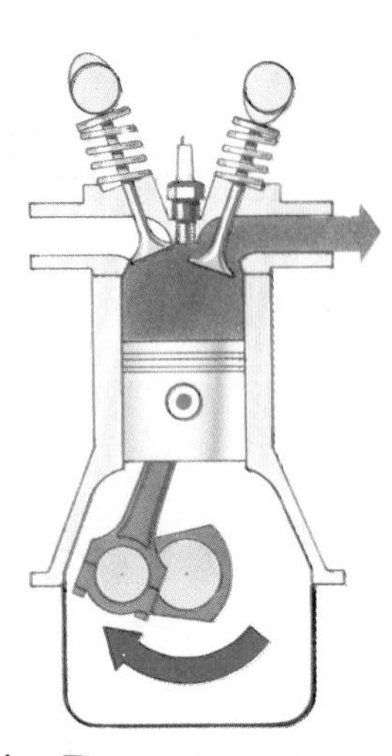

4. Burnt gases are pushed out by the piston.

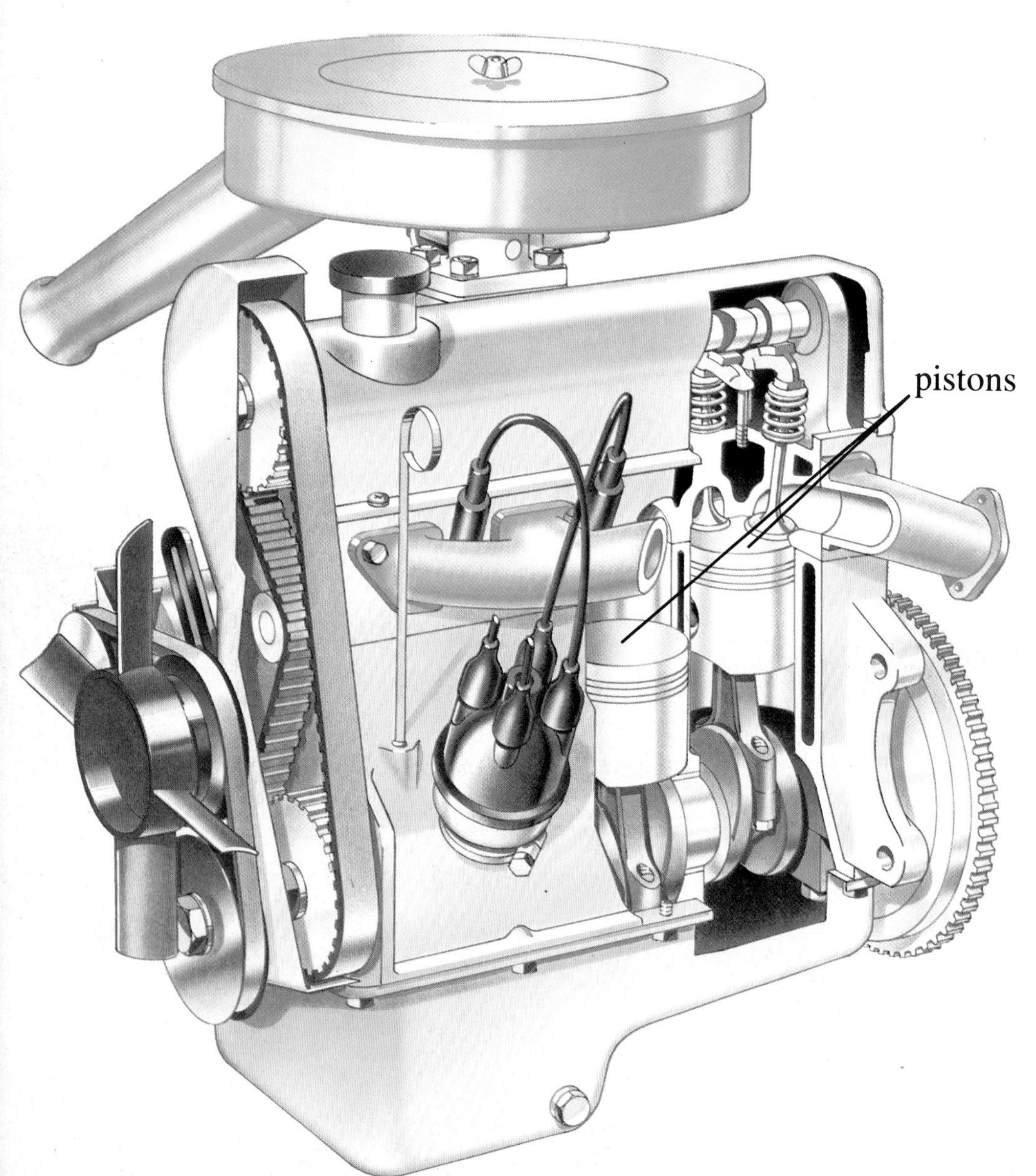

This engine is inside a giant ship. It is a diesel engine. Its inventor was Rudolf Diesel. It burns oil instead of gasoline. The oil burns when it meets hot air inside the cylinders. This makes the large engine run.

Turbines and jets

The turbine is one of the most powerful engines. It has turbine wheels inside. They are like this pinwheel. The sails turn when the boy blows through them. Turbine wheels turn when hot gas or steam passes through them. In jet planes, the gases rush out behind the engine to push it along.

Electricity is made in power stations. They use steam turbine engines. Steam in large boilers goes into the turbines. This makes them spin. Spinning turbines work generators. They make the electricity we all use.

row of generators

Electric motors

A magnet and coil of wire inside an electric motor make it work. Electricity is sent along this wire. This makes an electromagnet. When the coil of wire meets the magnet, it tries to push closer to it. Then it pulls away from it. This pushing and pulling runs the motor.

One end of the magnet is called the north. The other end is the south. They try to come together.

The two north ends push each other apart. So do the south ends.

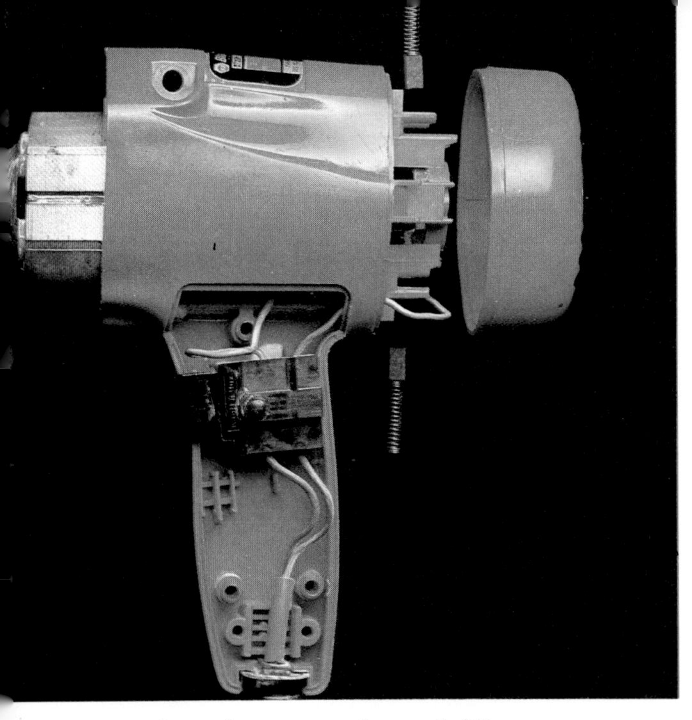

electric motor in a drill

Look below at the metal block with a rod through it. Wire is wound around each end of it. Electricity from a battery makes the wire and the block become magnets. The block magnet pushes and pulls. It spins between the horseshoe magnet. The spinning rod turns the gears and the wheels of the truck.

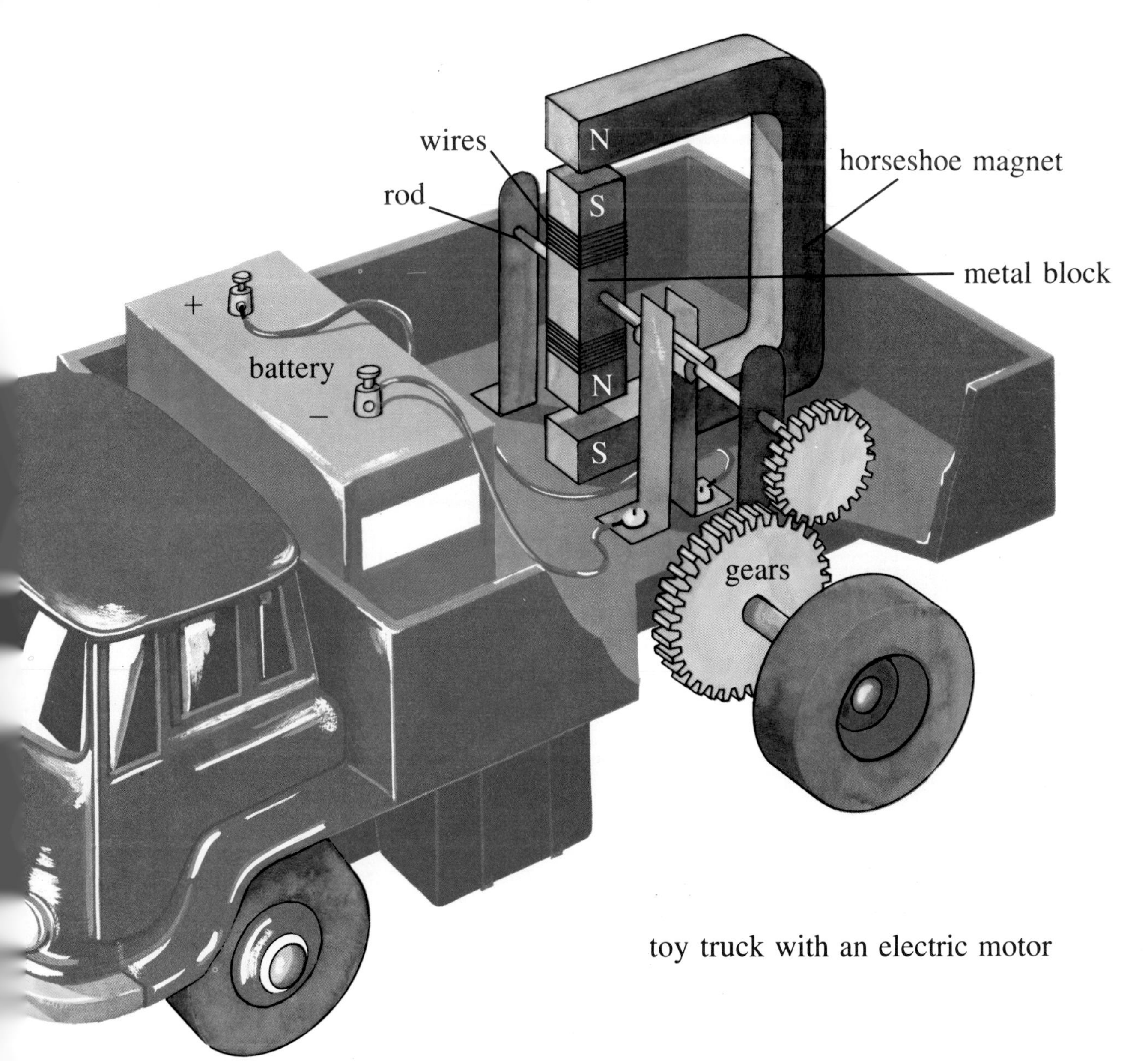

toy truck with an electric motor

GLOSSARY

These words are defined the way they are used in the book.

axle (AK sihl) a bar on which a wheel turns; cars and trucks have axles

battery (BAT ur ee) anything that stores electricity for later use; a car's battery stores electricity for use by its lights and starter

blade (blayd) a sharp, flat piece of metal used for cutting; a lawn mower has a large blade

block and tackle (blahk and TAK uhl) two pulleys, called blocks, held together by a rope; working together, they can lift very heavy loads

can opener (kan OH peh nur) a machine that cuts off the tops of cans to open them

chariot (CHAIR ee eht) a two-wheeled cart used long ago in wars; horses pulled the chariots

clock (klahk) a machine that shows what time it is

clockwork (KLAHK wurk) the parts of a clock that make it run

combine harvester (KAHM byn HAHR vihs tur) a large machine used for cutting crops and separating their seeds from the rest of the plant; corn, wheat, and other crops are harvested with combine harvesters

complicated machine (KAHM pleh *KAY* tihd muh SHEEN) a machine that is made up of two or more simple machines

conveyor (kuhn VAY ur) something that carries things from place to place; a conveyor belt can help move things around in a factory

crane (krayn) a machine with a long, swinging arm that lifts and moves heavy things

cycling (SY kling) riding a bicycle or motorcycle

cylinder (SIHL ehn dur) a long, round object; cylinders may be hollow or solid

diesel engine (DEE zihl EN jihn) an engine that burns oil as fuel

drill (drihl) a farm machine that plants seeds; also, the name of an electric machine tool

drive shaft (DRYV shaft) a bar that sends the movement of an engine to wheels to make them turn

electricity (ih LEK *TRIS* ih tee) a form of energy that gives us light and heat and runs many kinds of machines

elevator (*EL* ih VAY tur) a thing that lifts people and objects from a low place to a higher one

engine (EN jihn) a complicated machine used to make cars, trains, ships, trucks, tractors, and many other things move

escalator (*ES* kuh LAY tur) a moving stairway; people stand still while the escalator moves them up or down on it

excavator (*EKS* kuh VAY tur) a heavy machine used to dig large holes in the ground

factory (FAK tur ee) a building in which things are made; candy is made in a candy factory

fertilizer (*FUR* tih LY zur) any material that helps plants to grow

forage harvester (FAWR ij HAHR vihs tur) a farm machine that cuts grass and then forms it into bundles; this is winter feed for farm animals

furnace (FUR nihs) a thing used to produce heat; furnaces keep our homes warm in winter

gadget (GAJ it) any small object used to help get work done; a potato peeler is one kind of gadget

gasoline (GAS uh leen) a liquid chemical that catches fire very easily; it is used as fuel in cars, trucks, tractors, and airplanes

gasoline engine (GAS uh leen EN jihn) an engine that burns gasoline as fuel

gear wheels (GIHR weelz) wheels that have slots on their edges so they can fit against each other and turn at the same time

generator (*JEN* ur AY tur) a machine that makes electricity; turning parts of the machine generate, or make, electricity

grab excavator (grab *EKS* kuh VAY tur) an excavator that has a big scoop that grabs large amounts of soil

grinder (GRYN dur) a machine that chops up foods and other materials

harrow (HAIR oh) a farm machine used to break up plowed soil

harvest (HAHR vihst) to gather grain or other crops; also, the name given to all the crops gathered in a year

harvester (HAHR vihs tur) a farm machine used to gather grain and other crops

hedge trimmer (hehj TRIM ur) a machine used to trim, or shape, hedges and other bushes

jet (jeht) a kind of engine used on large airplanes; hot gases moving out the back of a jet engine move the airplane forward

kitchen (KIHCH ihn) a room in which foods are prepared; cooking is done in the kitchen of a home

lathe (laythuh) a machine tool for holding things and cutting them to a certain shape; chair legs are shaped by a lathe in a furniture factory

lever (LEHV ur) a bar used to lift or move a weight at one end by pushing down on the other end; a seesaw is a lever

machine (muh SHEEN) something that helps people do work; wheels and levers are simple machines

machine tool (muh SHEEN tool) a tool that runs with little or no help from people; an electric drill is a machine tool

millstones (*MIL* STOHNZ) round, flat stones that grind grain into flour; they are part of a mill whose wheels are driven by water or wind

mixer (MIHK sur) a machine used for mixing things; we can use a mixer to put flour and other things together to make a cake

moped (MOH pehd) a bicycle that has a motor

motor (MOH tur) a machine that makes other machines go; an engine

motorcycle (*MOH* tur SY kihl) a heavy two-wheeled machine that is run by an engine

mower (MOH ur) a machine that cuts grass and weeds

nutcracker (NUT *KRAK* ur) a simple machine used for breaking the shells of nuts

overhead crane (OH vur *HEHD* krayn) a crane that runs on rails high above the ground

pendulum (PEN juh luhm) a weight hanging from a fixed point so it can swing freely

photocopying machine (FOH toh *KOP* ee ing muh SHEEN) an office machine that takes pictures of printed things or drawings

pick (pihk) a pointed tool with

a handle used to break up hard soil; picks are used by ditch diggers

piston (PIS tuhn) a round piece of wood or metal that fits inside a cylinder, where it is forced by gas pressure to move back and forth, or up and down; pistons make a car engine run

pliers (PLY urz) a tool used to hold on to things and to bend wires

plow (plow) a farm machine that cuts and turns soil; also, a machine used for removing snow

power saw (POW ur saw) a large saw that is run by a motor

pruning shears (PROO ning shihrz) a tool used to cut away small branches on bushes and trees

pulley (PULL ee) a wheel and rope combination used to lift things easily; it is a simple machine

radiator (*RAY* dee AY tur) something that gets rid of heat by moving a cooling liquid through it; water moves through the radiator of a car to cool the car's engine

ramp (ramp) a sloping surface that connects a low place with a higher one

refrigerator (ree *FRIHJ* eh RAY tur) a machine that makes ice and keeps foods cold; refrigerators keep foods from spoiling

robot (ROH buht) a machine that does work that people usually do

roller (ROH lur) a cylinder of wood, metal, or stone used for smoothing, crushing, or pressing

screw (skroo) a kind of nail that has grooves that go all around it; screws are used to fasten metal or wood parts together

seesaw (see saw) a plank that rests on a support under its middle part; its ends can move up and down

shaper (SHAY pur) a machine tool used to give a certain shape to metal, wood, or plastic

shovel (SHUHV ehl) a simple machine used for digging

spiral (SPY rehl) anything that winds around and widens as it goes

staircase (*STAIR* KAYS) all the stairs, or steps, in one place as well as their supporting parts

steam engine (steem EN jihn) an engine that is run by steam power

steam turbine (steem TUHR bihn) a turbine engine run by steam power

teleprinter (TEL eh PRIHN tur) a machine that receives Telex messages and puts them on paper with a typewriter

Telex (TEL eks) a machine that sends messages from a typewriter over long distances

tongs (tongz) a tool with two narrow parts that are squeezed together to grab things; long tongs are used to put a log in a fireplace

tower crane (TOW ur krayn) a very tall crane used for lifting materials needed for putting up tall buildings

tractor (TRAK tur) a machine used by farmers, builders, and road makers

transmission (trans MISH shun) the part of a car that makes its wheels move; transmission gears can make the car go faster or slower

trolley (TRAHL ee) a kind of pulley that runs along a rail to help move things

turbine (TUHR bihn) a kind of engine that has wheels with vanes that turn when a gas or water pushes against them

typewriter (TYP *RY* tur) an office machine with alphabet letters on its keys; punching the keys makes words form on a sheet of paper

vacuum cleaner (VAK yoom KLEE nur) a machine used to clean carpets and other surfaces

vibrate (VY brayt) to move very fast back and forth in one place; the hood of a car vibrates when its engine is running

water wheel (WAHT ur weel) a wheel moved by flowing water; water wheels were once used to run machines

wedge (wehj) a narrow piece of metal or wood that is thick at one end and thin at the other end; a wedge is pounded into a log to split it apart

wheels (weelz) round objects that are used to roll whatever they are attached to; wheels are simple machines

windmill (*WIND* MIL) a machine run by wind; a windmill hooked to a pump can pull water from a well

wrench (rench) a tool used to turn pipes and other things

FURTHER READING

Ackins, Ralph. *Energy Machines*. Milwaukee: Raintree Publishers, 1980. 32pp.

Alth, Max. *Motorcycles and Motorcycling*. New York: F. Watts, Inc., 1979. 90pp.

Aylesworth, Thomas G., ed. *It Works Like This*. New York: Natural History Press, 1968.

Buehr, Walter. *First Book of Machines*. New York: F. Watts, Inc., 1962.

Dolan, Edward. *Engines Work Like This*. New York: McGraw-Hill Book Co., 1972.

Fisher, Leonard E. *Glassmakers*. New York: F. Watts, Inc., 1964.

Fleming, Alice. *Wheels*. New York: J. B. Lippincott Company, 1960. 176pp.

Girard, Pat. *Flying Machines*. Milwaukee: Raintree Publishers, 1980. 32pp.

Hancock, Ralph. *Super Machines*. New York: Viking Press, 1978.

Hellman, Hal. *The Lever and the Pulley*. New York: M. Evans and Company, Inc., 1971. 45pp.

Howard, Sam. *Communications Machines*. Milwaukee: Raintree Publishers, 1980. 32pp.

James, Elizabeth and Carol Barkin. *The Simple Facts of Simple Machines*. New York: Lothrop, Lee and Shepard Company, 1975. 64pp.

Kiley, Denise. *Biggest Machines*. Milwaukee: Raintree Publishers, 1980. 31pp.

Liberty, Gene. *The First Book of Tools*. New York: F. Watts, Inc., 1960. 62pp.

Lord, Beman. *Look at Cars*. rev. ed. New

York: Walck, 1970. 48pp.
Pick, Christopher C. *Oil Machines*. Milwaukee: Raintree Publishers, 1979. 31pp.
Pine, Tillie; Levine, Joseph. *Simple Machines and How We Use Them*. New York: McGraw-Hill Book Company, 1965.
Stone, William D. *Earth Moving Machines*. Milwaukee: Raintree Publishers, 1979, 31pp.
Weiss, Harvey. *Motors and Engines and How They Work*. New York: T. Y. Crowell, 1969.
Wykeham, Nicholas. *Farm Machines*. Milwaukee: Raintree Publishers, 1979. 31pp.

QUESTIONS TO THINK ABOUT

Machines All Around

Do you remember?

How did people do their work before they had machines?

When was the wheel invented?

What do pulleys do?

How do two gear wheels that are touching move?

What kind of simple machine is a shovel?

Find out about . . .

Moving loads without wheels. Long ago the Indians who lived on the American continent had to move their belongings from place to place without the help of wheels. How did they do this? How did the Eskimos move their loads across ice and snow? What did ancient Egyptians use to move large stone blocks to build pyramids?

The kinds of pulleys. How many are there? How big are the biggest ones? What are all these pulleys used for? Could our industries get along without them?

In the Home

Do you remember?

Name several uses of machines in the kitchen.

How does a vacuum cleaner work?

What are pliers used for?

What two kinds of simple machines can you find in a vice?

What does the word clockwork mean?

What do hanging weights do for a clock?

Name three kinds of machines that are used in the yard.

Find out about . . .

Laundromats. What are they? What is it like inside a laundromat? What do its machines do? What makes them go? How do they compare with laundry machines used in homes?

Machines used in camping. What kinds of tools and machines do campers use? How can pulleys help them? Is any kind of wedge used? If so, how is it used? In what ways could wheels and engines help campers?

On the Farm

Do you remember?

In what three ways do machines help farmers produce food?

What machine now does the farm work that horses used to do?

What is the name of the machine farmers use to sow seeds?

What do the blades of a plow do?

What are the two kinds of engines that farm tractors may have?

What is a harvester?

How does a potato picker work?

Find out about . . .

Old tractors. How big were the earliest kinds of tractors? How were they powered? Could the owner of a small farm afford to buy one? Why did some of them look like locomotives? What did they use for fuel?

Small farm machines. How are machines used to fix fences and farm buildings? What kinds might be used to fix tractors? How do farmers prepare animal feeds? Are special kinds of machines used on dairy farms?

In the Town

Do you remember?

How do heavy machines help build our big buildings?

What does a crane look like?

What is a tower crane used for?

What is an excavator?

How does a grab excavator work?

How does a road drill work?

What is an escalator used for?

What is a Telex machine?

Name three kinds of office machines.

Find out about . . .

A modern office. How many different kinds of machines are used in offices? How many of these are simple machines? How many of them are run by electricity?

Skyscrapers. How do people get to their offices at the top of tall buildings? How long does it take to go up or down? When was the first skyscraper built? Could our tallest skyscrapers have been put up before elevators and escalators were invented?

In the Factory

Do you remember?

How are modern factories different from the factories of long ago?

Name two kinds of machines that are used in a furniture factory.

What is a conveyor belt used for?

What is a factory robot?

What is a machine tool?

Machine tools do five kinds of work. Name three of them.

Find out about . . .

Kinds of wineglasses. How many kinds of glasses are used just for wine drinking? What kinds of shapes do they have? How big is the biggest one? How small is the smallest one? Do you think they are all made the same way?

Lathes. What kinds of lathes are there? What kind of work do they do? Can robots run them? How much training must people have to run them?

On the Road

Do you remember?

How fast can a racing bicycle go?

What is a moped?

How is a motorcycle's engine started?

How many parts does a car have?

How does the radiator help a car's engine?

What makes a car's wheels turn around?

What does the battery do for a car?

Find out about . . .

Ancient cars. How were early cars different from cars today? What kind of fuel did they use? How fast did they go? Were their drivers comfortable in all kinds of weather? What kind of tires did they have?

Early motorcycles. What did they look like? How fast did they go? How far could they travel on a gallon of fuel? Were they used mostly for fun?

Power

Do you remember?

How did heavy work get done before there were engines to drive machines?

Why are donkeys used for hauling heavy loads in places where the land is rough and steep?

What two main jobs can windmills do?

Who was George Stephenson?

Why does a steam train need water?

What does a diesel engine burn for fuel?

What do power stations use steam turbine engines for?

Find out about . . .

Magnets. How many different kinds are there? What are they used for? How is an electromagnet different from other kinds? What is it used for?

Rudolf Diesel. Where was he born? How did he get the idea for his new kind of engine? What kinds of industry make good use of his engine?

PROJECTS

Project — My Automobile Book

Get six pieces of drawing paper. They should be the large size. Fold each sheet in half. Now hold one sheet in your left hand. Hold it by the folded edge. Then slide the rest of the folded sheets into the one you are holding. If you did this right, you should be able to open and turn the sheets as if they were pages of this book.

Next, using a black crayon, number the pages from 1 to 24. Write the numbers near the bottom of the sheets. They should look like the page numbers in this book. Now, print near the top of the first page, ''My Automobile Book.''

For the next step in this project you will need a pair of scissors and some glue or paste. You will also need some old magazines and newspapers that you can cut up. Look through them for pictures of different kinds of automobiles. With a pencil, write the name of the car on any picture you decide to cut out. Cut out cars of all different sizes and models.

Look over all your cutouts. Decide which ones you should paste on page one of your book. Arrange them the way you like them best. Then paste them on the page. Then write the name of the car under the picture. Do the same thing for all the rest of the pages. If you can't find enough pictures, make large drawings of cars you like. Paste your drawings on the blank pages that are left.

Project — Building the Pyramids

Look up about the pyramids in an encyclopedia. Read all you can about how they were built by ancient Egyptians. Ask your librarian for children's books about the pyramids. Look carefully at the pictures. Look at the tools and machines the workers used. Make small sketches of them so you will remember what they looked like.

Now you are ready to put what you have learned to work. Get a piece of posterboard and some crayons. Draw a scene on the posterboard that will show how the pyramids were built. Draw all the tools and machines you think the workers used. Did they use pulleys and ropes? Did they use rollers to move big stones? Were any ramps used?

Project — Airplane Engines

Make a two-column chart showing the difference between jet engines and propeller-driven airplane engines. Draw pictures or use words to show this.

INDEX

Photo Credits:
Chris Barker; Paul Brierly; Camera Press; J. Allan Cash; Coles Cranes; Doxford Engines; Dunlop-Angus; Mary Evans; Massey Ferguson; Ingersol-Rand Co; Robin Kerrod; Herbert Morris; G. R. Roberts; Rolls Royce; Seiko; Theorem; Thomas Wilkie; Woolf Tools; Zefa Picture Library; Front Cover: Allis-Chalmers Corp.